Petroleum and Gas Field Processing

Petroleum and Gas Field Processing

Editor

Manoj Karkare

Petroleum and Gas Field Processing

Edited by **Manoj Karkare**

Printed in 2017

ISBN: 978-1-68117-429-7

Library of Congress Control Number: 2015936543

© 2016 by
SCITUS Academics LLC,
616, Corporate Way, Suite 2, 4766,
Valley Cottage, NY 10989

www.scitusacademics.com

This book contains information obtained from highly regarded resources. Copyright for individual articles remains with the authors as indicated. All chapters are distributed under the terms of the Creative Commons Attribution License, which permits unrestricted use, distribution, and reproduction in any medium, provided the original author and source are credited.

Notice

Reasonable efforts have been made to publish reliable data and views articulated in the chapters are those of the individual contributors, and not necessarily those of the editors or publishers. Editors or publishers are not responsible for the accuracy of the information in the published chapters or consequences of their use. The publisher believes no responsibility for any damage or grievance to the persons or property arising out of the use of any materials, instructions, methods or thoughts in the book. The editors and the publisher have attempted to trace the copyright holders of all material reproduced in this publication and apologize to copyright holders if permission has not been obtained. If any copyright holder has not been acknowledged, please write to us so we may rectify.

Contents

Preface vii

Chapter 1 Analysis of Petroleum System for Exploration and Risk Reduction in Abu Madi/Elqar'a Gas Field, Nile Delta, Egypt 1
Said Keshta, Farouk J. Metwalli, and H. S. Al Arabi

Chapter 2 Research of Drainage Gas Recovery Technology in Gas Wells 23
Shuren Yang, Di Xu, Lili Liu, Chao Duan, and Liqun Xiu

Chapter 3 Alternative System of Industrial Paint Applied to Spherical Mount for Liquefied Petroleum Gas 41
Fernando B. Mainier, Francisco Otavio Pereira da Silva, Gilberto Oliveira da Silva

Chapter 4 Life-Cycle Analyses of Energy Consumption and GHG Emissions of Natural Gas-Based Alternative Vehicle Fuels in China 55
Xunmin Ou and Xiliang Zhang

Chapter 5 Study on Nonequilibrium Effect of Condensate Gas Reservoir with Gaseous Water under HT and HP Condition 77
Dali Hou, Pingya Luo, Lei Sun, Yong Tang, and Yi Pan

Chapter 6 Experimental and Modeling Studies on the Prediction of Gas Hydrate Formation 101
Jian-Yi Liu, Jing Zhang, Yan-Li Liu, Xiao-Hua Tan, and Jie Zhang

Chapter 7 Microbial Degradation of Petroleum Hydrocarbon Contaminants: An Overview 115
Nilanjana Das and Preethy Chandran

Chapter 8 Pore Structure and Limit Pressure of Gas Slippage Effect in Tight Sandstone 149
Lijun You, Kunlin Xue, Yili Kang, Yi Liao, and Lie Kong

Chapter 9	**Methods for Separation, Recycling and Reuse of Biodegradation Products** 169	
	Ganapati D. Yadav and Jyoti B. Sontakke	

Citations 217

Index 221

Preface

In this book, we discuss about the petroleum and gas field processing. Oil and gas wells consist of mixtures of oil, gas, and water that are difficult to transport, requiring a certain amount of field processing. This reference analyzes principles and procedures related to the processing of reservoir fluids for the separation, handling, treatment, and production of quality petroleum oil and gas products. It details strategies in equipment selection and system design, field development and operation, and process simulation and control to increase plant productivity and safety and avoid losses during purification, treatment, storage, and export. Providing guidelines for developing efficient and economical treatment systems, the book features solved design examples that demonstrate the application of developed design equations as well as review problems and exercises of key engineering concepts in petroleum field development and operation.

Editor

Chapter 1

Analysis of Petroleum System for Exploration and Risk Reduction in Abu Madi/Elqar'a Gas Field, Nile Delta, Egypt

Said Keshta[1], Farouk J. Metwalli[2], and H. S. Al Arabi[3]

[1]Geology Department, Faculty of Education, Suez Canal University, Arish, Egypt
[2]Geology Department, Faculty of Science, Helwan University, Cairo, Egypt
[3]Geology Department, Faculty of Science, Suez Canal University, Ismailia, Egypt

ABSTRACT

Abu Madi/El Qar'a is a giant field located in the north eastern part of Nile Delta and is an important hydrocarbon province in Egypt, but the origin of hydrocarbons and their migration are not fully understood. In this paper, organic matter content, type, and maturity of source rocks have been evaluated and integrated with the results of basin modeling to improve our understanding of burial history and timing of hydrocarbon generation. Modeling of the empirical data of source rock suggests that the Abu Madi formation entered the oil in the middle to upper Miocene, while the Sidi Salem formation entered the oil window in the lower Miocene. Charge risks increase in the deeper basin megasequences in which migration hydrocarbons must traverse the basin updip. The migration pathways were principally lateral ramps and faults which enabled migration into the shallower middle to upper Miocene reservoirs. Basin modeling that incorporated an analysis of the petroleum system in the Abu Madi/El Qar'a field can help guide the next exploration phase, while oil exploration is now focused along post-late Miocene migration paths. These results suggest that deeper sections may have reservoirs charged with significant unrealized gas potential.

INTRODUCTION

The Nile Delta basin contains a thick sequence of potential hydrocarbon source rocks that generate essentially gas and condensate.

The Nile Delta is generally known as a natural gas-prone (essentially methane/gas condensate) region with production from Miocene and Pliocene fields. However, temperature and pressure data from these fields suggest that the Nile Delta basin should be oil rather than gas prone [1].

The purpose of this paper is to evaluate potential source rocks in north eastern part of Nile Delta from available geochemical data from two wells. An additional objective is to study the regional petroleum systems by using numerical models which provide information about burial and temperature history, maturation of source rocks and timing of expulsion of hydrocarbons. Maturity information was used for calibration of numerical models.

Sharaf [2] showed from organic geochemical and petrographic analyses that the kerogen in the early Pliocene mudstones of the Kafr El-Sheikh formation and the Tortonian Wakar formation in the NE of the Nile Delta has poor capability to generate gas and minor oil. These formations are immature in all of the wells he studied. However, the early Miocene Qantara and the Middle Miocene Sidi Salem formations have a poor potential to generate gas and minor oil in the southern part of the area, further north Sharaf [2]. They have improved capability to generate oil and minor gas. These formations are immature in the southern parts of the area but are within the oil zone in the northern part [2].

In general, the kerogen in the Pliocene and Middle-Late Miocene samples from the NE Nile Delta is mainly of Continental origin. Terrestrial woody and herbaceous fragments [2] are the main components with a minor content of amorphous organic matter (AOM) and marine phytoplankton. In the Early Miocene and Oligocene samples from above 3500 m, the kerogen is mainly of Terrestrial origin. Below this depth, the kerogen quality improves with increasing depth (the content of AOM increases to 75% and the Terrestrial fragments decrease to 20%). The Eocene to late Cretaceous samples are characterized by moderate to high AOM and marine phytoplankton contents typical of kerogen with a good to fair petroleum generation potential. In the early-middle Cretaceous and late Jurassic samples, the kerogen is an amorphous-woody-algal assemblage accompanied by significant proportion of inertinitic debris [2].

THE AREA OF STUDY

The area of study covers the onshore concession within Nile Delta. It lies between latitudes 31°20' and 31°35'N and longitudes 31°15' and 31°30'E (Figure 1). A total of 30 wells were drilled, of which 21 are in Abu Madi lease and 9 wells are in the El Qar'a lease. Twenty-four wells gave positive results and six wells were plugged and abandoned as dry holes, (five wells in Abu Madi lease and one well in El Qar'a lease (Figure 1). As a matter of fact, Abu Madi field is the first commercial discovery in the Nile Delta, where the IEOC achieved the first gas discovery from the early pliocene Abu Madi formation in the northeastern part of the onshore Delta by drilling the Abu madi-1 well, encounteringtwo pay zones with more than 50 m thick gas and condensate bearing sands in Abu Madi formation [4]. The field was put on production in 1975 and was producing 355 MMSCF per day in 1995 [4].

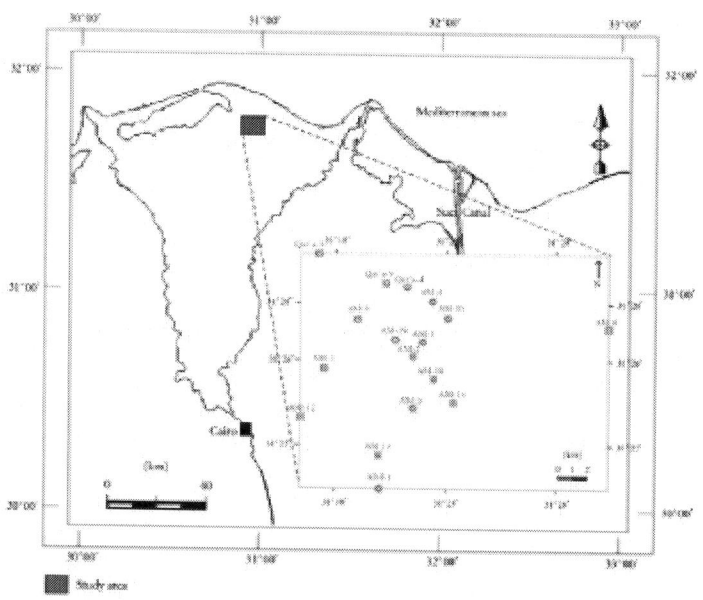

Figure 1: Location map of the study area.

GEOLOGICAL SETTING

The sedimentary section in the Nile Delta area with gas potential seems to be limited to the Neogene formations trapped against listric faults or draped over tilted fault blocks. However pre-Miocene formations of the base of this Neogene sequence may also be considered as future exploration plays. Mesozoic reservoirs are present at greater depth and have been only penetrated by a few wells which are mostly located in the south delta block.

The sedimentary rocks penetrated in Abu Madi/El Qar'a field consist of thick clastics representing Miocene-Holocene time interval. These rocks were described by El Heiny and Enani [3], Figure 2, Alfy et al. [5], and Sarhan and Hemdan [6].

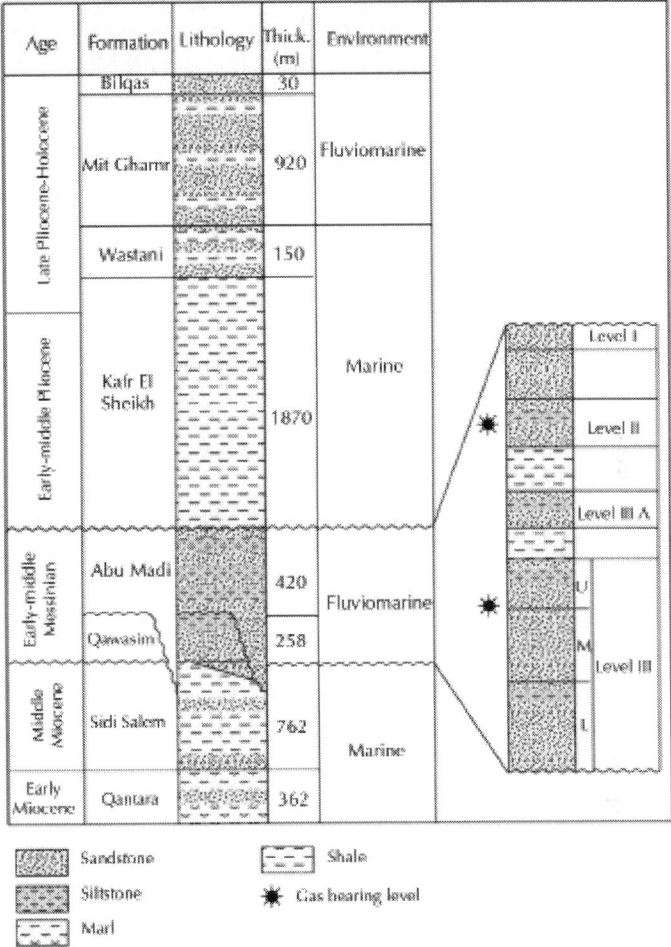

Figure 2: Generalized lithostratigraphic column of Abu Madi gas field modified after [3].

The studied section involves lithostratigraphic units ranging in age from Miocene to Holocene. The studied section is differentiated into the rock units: Qantara, Sidi Salem, Qawasim, and Abu Madi formations of the Miocene age; Kafr El Sheikh and El Wastani formations of the Pliocene age; Mit Ghamer formation of late Pliocene-Pleistocene age and the Bilqas formation of recent age. All these Formations consist essentially of clastic sediments (shale, sand, and silt).

MATERIALS AND METHODS

One directional modeling of burial history and thermal maturity was preformed on five wells using basin mod Platter River (2003) software in addition to kerogen data measured by Khaled et al. [7], these samples were taken from Abu Madi formation and they represent a depth interval from 3067 m to 3390 m in well Abu Madi-9 and, 3108 m to 3460 m in Abu Madi-11 well (Tables 1 and 2).Parameters measured include TOC, S1, S2, S3 and temperature of maximum pyrolysis yield (T_{max}). Hydrogen index and oxygen index were calculated as described by Espitalie et al. [8] and peters and Cassa [9].

Table 1: Summary of pyrolysis data of Abu Madi-9 well

Depth (M)	TOC	S1	S2	S3	Tmax	HI	OI	S2/S3	%RO	TTI
3067	0.68	0.15	0.51	0.85	427	75	125	0.6	0.6	9
3154	1.21	0.17	2.74	1.1	429	226	91	2.49	0.62	12
3184	1.12	0.14	2.1	0.95	429	187	85	2.21	0.6	9
3224	1.37	0.1	1.06	1.01	431	77	74	1.05	0.5	3
3256	1.54	0.05	1.1	1.06	434	72	69	1.04	0.7	20
3310	1.85	0.1	1.74	0.7	434	94	38	2.49	1.1	98
3344	0.81	0.09	0.71	0.25	436	87	31	2.84	1.3	165
3358	0.58	0.13	0.63	0.24	437	108	42	2.63	0.9	48
3390	0.51	0.25	0.25	0.17	439	49	34	1.47	1.35	179

Table 2: Summary of pyrolysis data of Abu Madi-11 well

Depth (M)	TOC	S1	S2	S3	Tmax	HI	OI	S2/S3	%RO	TTI
3108	1.5	0.07	1.38	1.95	429	89	125	0.71	0.49	0.8
3186	1.26	0.1	1.27	1.41	430	101	112	0.9	0.55	7
3210	1.65	0.06	1.42	1.7	431	86	103	0.84	0.71	21
3231	1.57	0.09	1.54	1.56	432	98	99	0.99	0.67	18
3250	1.62	0.11	1.41	1.07	432	87	66	1.32	0.88	44
3293	1.27	0.2	1.42	1.04	436	112	82	1.37	0.7	20
3318	0.82	0.05	0.75	0.77	435	91	94	0.97	0.68	18
3400	1.66	0.08	1.61	0.86	435	97	52	1.87	1.1	99
3425	2.03	0.1	0.91	1.02	436	45	50	0.89	1.15	112

| 3460 | 1.74 | 0.14 | 1.53 | 0.84 | 436 | 88 | 48 | 1.82 | 1.3 | 165 |

MODELING PROCEDURES

To Construct the burial history, the essential input data included formation tops from ground level, absolute time of deposition in Ma (millions of years), lithological composition, hiatus age, thickness and age of eroded intervals, and the heat flow data calculated from observed geothermal gradients. The study area had sufficient information for modeling. Absolute age in many of the different stratigraphic units was defined using the global stratigraphic chart complied by Cowie and Bassett [10]. The lithological composition of the stratigraphic units was obtained from composite logs, whereas the average porosities and densities of these reservoir units were determined from petrophysical analysis.

In the basin modeling, the sedimentary sequence at the wells is subdivided into layers from the Tortonian (12 Ma) to the present and have vertical continuity in lithology and lateral continuity in time. The vertical continuity is essential to correctly compute pressure and temperature histories. The lateral continuity in time is needed to accurately define chronostratigraphy and to plot the result at any time during the geologic history [11].

SOURCE ROCK EVALUATION

TOC and Pyrolysis Data

Total organic carbon (TOC) of samples from Abu Madi formation (Tables 1 and 2) ranges from 0.51 to 2.03 Wt% which is considered good potential source rock. The migration index (S1/TOC) ranges from 0.03 to 0.49 mg HC/g TOC with an average value of 0.11 mg HC/g TOC. This value lies in the range of 01–0.2 mg HC/g TOC suggested by Hunt [13] to characterize the oil expulsion zone. In the Rock-Eval pyrolysis analysis, free hydrocarbons (S1) in the rock and the amount of hydrocarbons (S2) and CO_2 (S3) expelled from pyrolysis of kerogen are measured (Tables 1 and 2). In addition, the T_{max} value, which represents the temperature at the point where the S2 peak is maximum, is also determined [14]. Pyrolysis data from 19 samples from Abu Madi Formation presents low values of S1 (average 0.12 mg HC/g rock), S2 value ranges from 0.25 to 2.74 mg HC/g rock (average is 1.27 mg HC/g rock), the (S1 + S2) range from 0.50 to 2.91 mg HC/g rock (average 1.38 mg HC/g rock). The calculated S2/S3 equals 1.3 (less than 2).These values indicate gas-prone organic matter and poor to fair hydrocarbon generation potential. The values of Production Index (PI) expressed by {S1/ (S1+S2)} range from 0.04 to 0.5 (average 0.11) lie in the range of oil window [13]. Accordingly, organic matter in mudstone bed within Abu Madi formation is suggested to be fairly mature and gas prone; they have reached the early oil generation zone very close to the roof of the oil window.

Vitrinite Reflectance

The vitrinite reflectance R_o values range from 0.5 to 1% indicating that the samples are thermally mature and have entered the mature to late mature stage of hydrocarbon generation. This is supported by pyrolysis T_{max} values as shown in Figure 3.

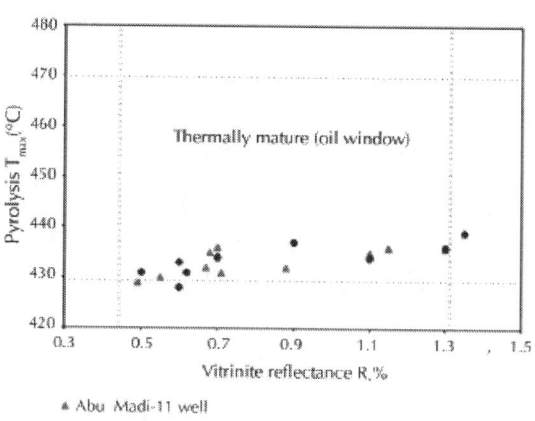

Figure 3: Plot of T_{max} versus Vitrinite reflectance values (R_o), showing good agreement between T_{max} and Vitrinite reflectance data.

Type of Organic Matter

The stage of maturity can be estimated using the temperature of maximum pyrolysis yield (T_{max}), although this is partly dependent on other factors such as the type of organic matter or mineral matrix effects [9]. In general, T_{max} values less than 435°C indicate immature organic matter, whereas values of about 445°C indicate the end of the oil window and the beginning of the wet gas zone [8]. A plot of T_{max} versus hydrogen index from Abu Madi-9 and Abu Madi-11 wells is shown in Figure 4. The maturity stages determined from this plot (Figure 4) indicate that the kerogen of mature type agrees with that determined from vitrinite reflectance.

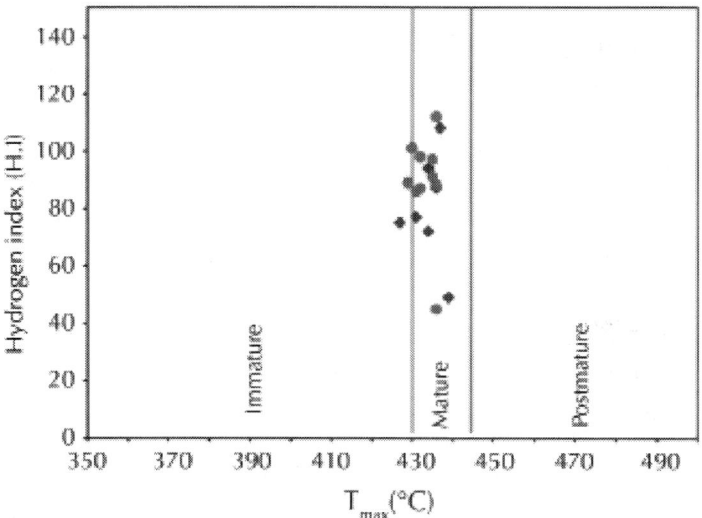

- Abu Madi-9 well data
- Abu Madi-1 well data

Figure 4: Plot of T_{max} versus H.I. for the analyzed samples from Abu Madi-9 and Abu Madi-11 wells well.

The hydrogen index (HI = S2/TOC) versus oxygen index (OI = S3/TOC) plot on the Van Krevelen diagram [15,16], for the data from Abu Madi-9 well is shown in Figure 5. The plot indicates the predominance of organic matter of type III.

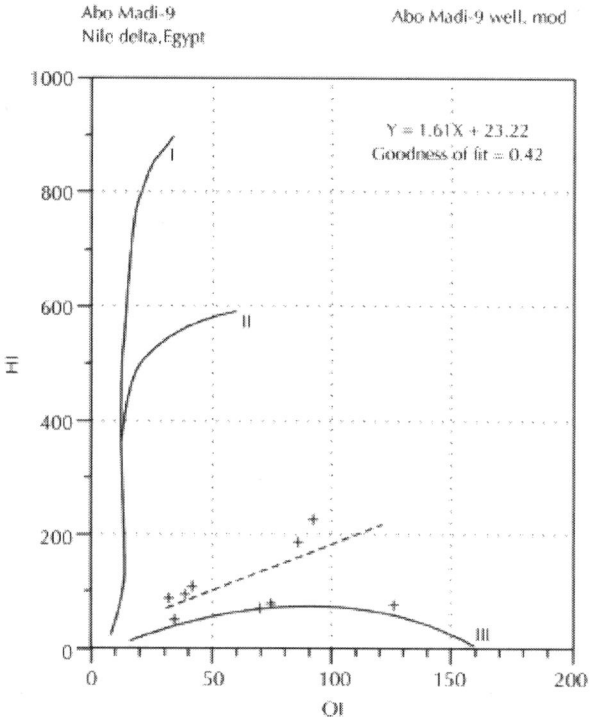

Figure 5: Plot of Oxygen index versus hydrogen index for the analyzed samples from Abu Madi-9 well.

Thermal Maturity of Organic Matter

The heat flow history of a basin is proposed by establishing an agreement between a calculated (or modeled) maturity parameter and the equivalent observed maturity parameter (such as vitrinite reflectance or Rock-Eval T_{max}). The maturity profile (Figure 6) reveals a good match between the measured and calculated vitrinite reflectance values. Windows boundary of oil and gas varies with type of organic matter, ranging from 0.5% to 1.0% R_0 and 1.3–3.5 R_0 respectively [14, 16]. Generally, vitrinite reflectance values increase with depth due to an increase in temperature and age of the rock with depth. The available data generally suggest that the majority of the kerogen of Abu Madi formation belong to mature type III in the principal zone of oil generation (oil window), where R_0 values range from 0.5 to 1%, with small amounts of kerogen immature type III, where R_0 is less than 0.5%. Figure 3 shows distribution of vitrinite reflectance data indicating that the samples taken from late Miocene Abu Madi formation are sufficiently thermally mature (oil window) for hydrocarbon generation (Figure 3).

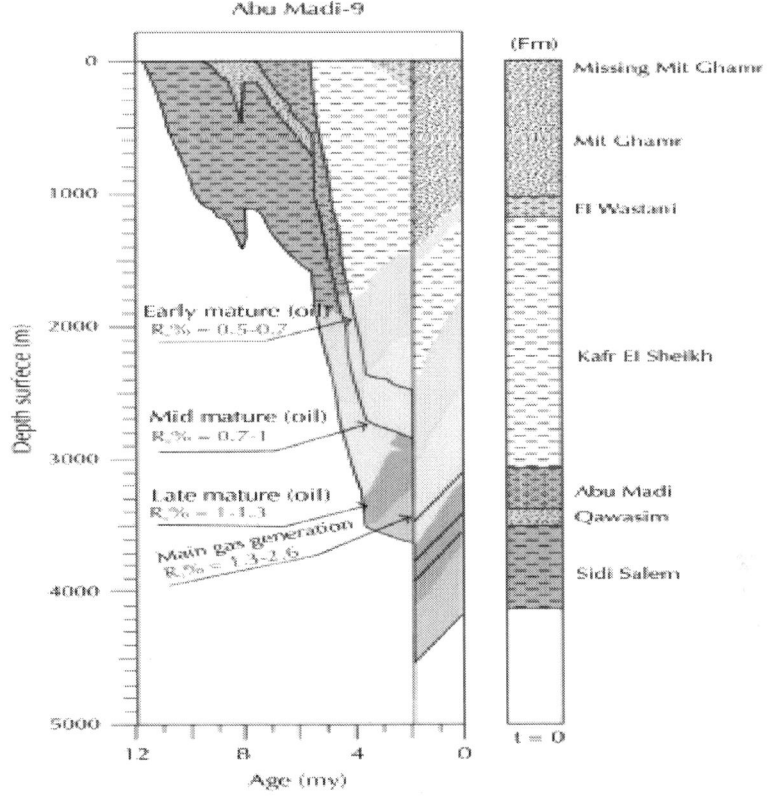

(a)

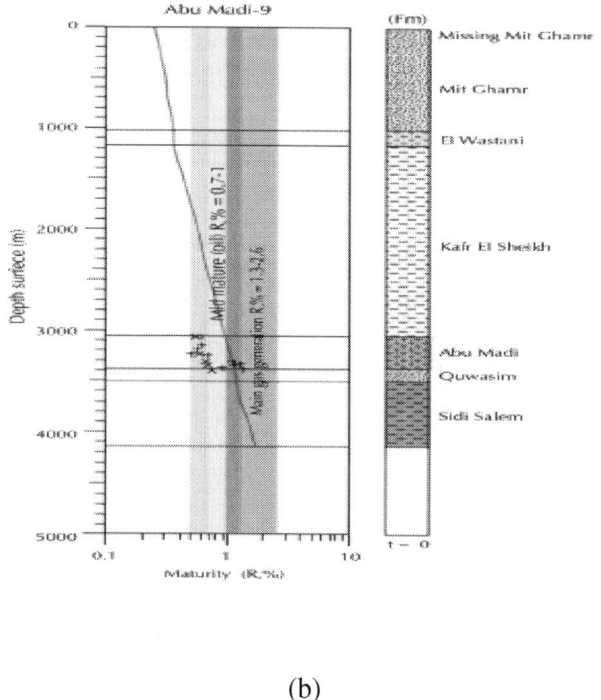

(b)

Figure 6: (a) Burial history curve for Abu Madi-9 well showing the calculated maturity (b) the measured maturity for samples taken from Abu Madi formation. Both figures agree the level of maturity for Abu Madi formation (early to mid-mature).

NUMERICAL MODELING

Burial History

Supporting evidence for the existence of prolific petroleum systems in the Abu Madi/El Qar'a gas field area comes from basin modeling in the area, which incorporated an analysis of the petroleum system criticals [17]. For effective exploration, a better understanding of the processes that led to the generation, migration, and accumulation of hydrocarbons is necessary. The purpose of the modeling is to evaluate the maturity of the potential source rocks and to estimate their timing of generation and expulsion. The calculated maturities of the potential source rocks were calibrated against the available measured maturity parameters, which included mainly vitrinite reflectance data (%R_o).

The burial history curves of two wells penetrating the study area are shown in Figures 7 and 8. Theses curves show the subsidence history of Abu Madi-09 and Abu Madi-15 wells including both steady state and rifting where the Basin mod software uses two basic assumptions for heat flow histories. The steady state uses a constant heat flow over time, whereas the nonsteady state (rifting) uses a variable heat flow over time. These curves (Figures 7 and 8) reveal three distinct pluses of subsidence and uplift. These pluses are from 12 to 5.5 Ma, from 5.5 to 3.5 Ma and from 3.5 Ma to the present.

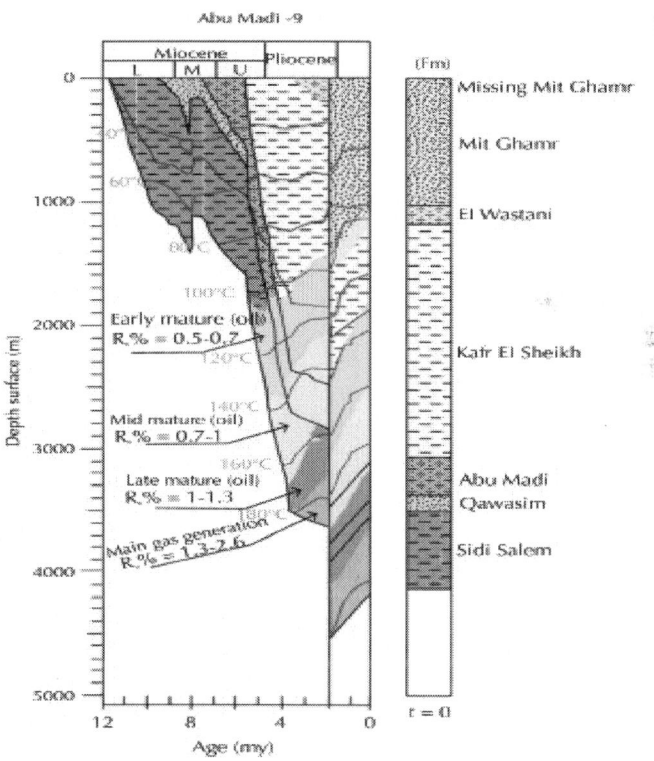

Figure 7: Burial history curve for Abu Madi-9 well showing kinetic window of stratigraphic units.

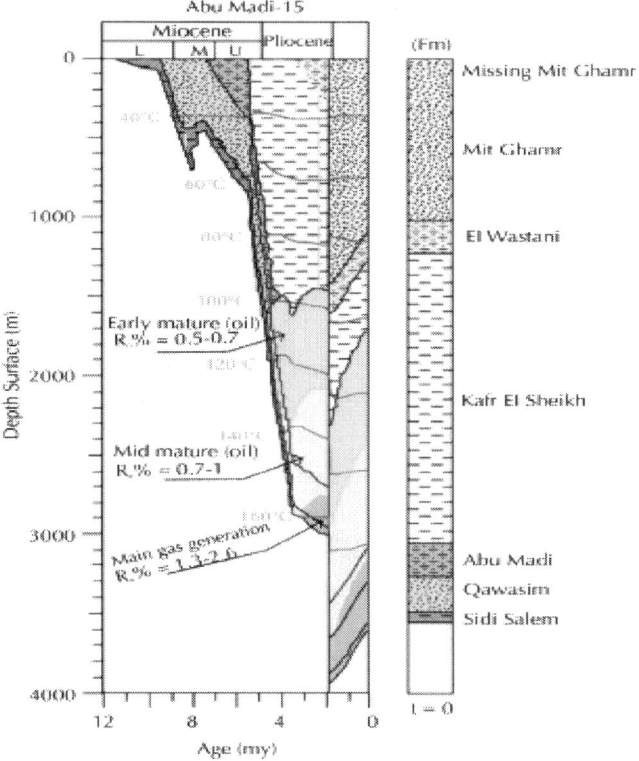

Figure 8: Burial history curve for Abu Madi-15 well showing kinetic window of stratigraphic units.

The burial history in Abu Madi/El Qar'a field is characterized by a relatively low rate of sediment accumulation from 12 to 5.5 Ma in the middle to upper Miocene (Tortonian-Messinian), during the deposition of Sidi Salem, Qawasim, and Abu Madi formations. A relatively brief period of uplift and erosion occurred between 8.5 and 7.5 Ma and it was ended with deposition of Abu Madi formation.

The rate of deposition increased substantially between 5.5 and 3.5 Ma in the lower Pliocene during deposition of Kafr El Sheikh formation, when most of the sediments (between 1810 ft to 2003 ft) were deposited. A slowing rate of sedimentation in the upper Pliocene occurred during deposition of El Wastani formation. This is followed by more rapid, post upper Pliocene, burial from 1.9 Ma to recent during deposition of Mit Ghamer formation.

Tectonic Subsidence Curves

The tectonic subsidence curves of the two selected wells are shown in Figures 9 and 10. These illustrate the tectonic subsidence and sedimentation rate in the Abu Madi/El Qar'a field area. The curves clearly show the period of non-deposition that occurred in the Middle Miocene, prior to deposition of the Abu Madi Formation (from 8.5 to 7.5 Ma). The curves also reveal the existence of two periods marked by an increase of subsidence/sedimentation rate, during deposition of Kafr El Sheikh and Mit Ghamer Formations.

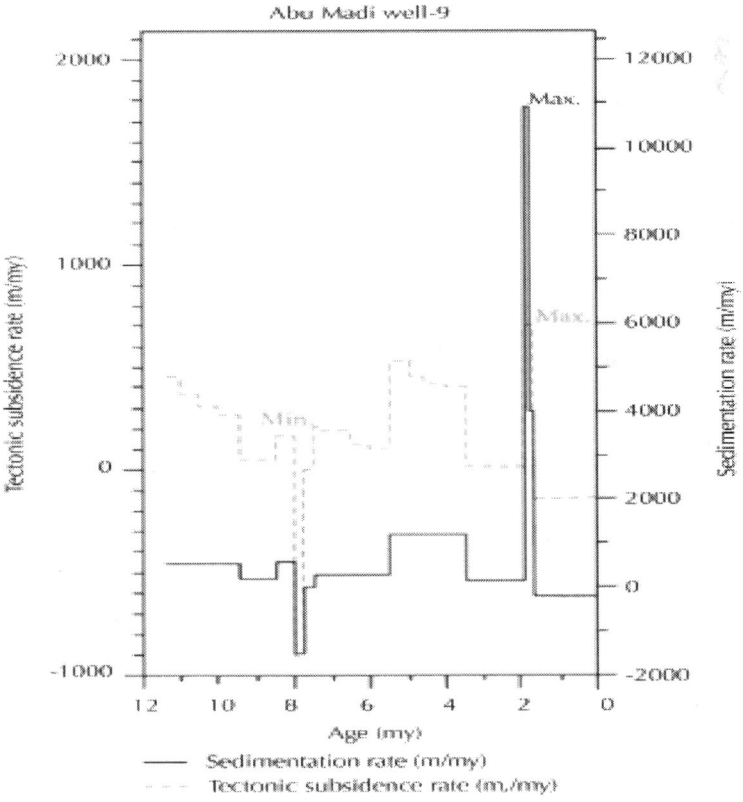

Figure 9: Tectonic subsidence rate for Abu Madi-9 well, showing period of nondeposition and period of increasing rate of subsidence/sedimentation.

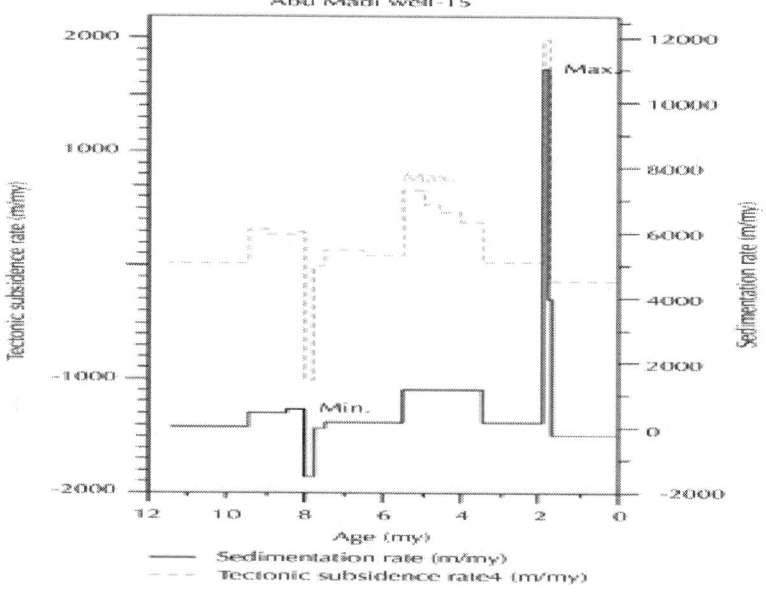

Figure 10: Tectonic subsidence rate for Abu Madi-15 well, showing period of nondeposition and period of increasing rate of subsidence/sedimentation.

Timing of Hydrocarbon Generation of Miocene Source Rocks

The timing of hydrocarbon generation from the Miocene Abu Madi and Sidi Salem formations in the studied wells was analyzed and determined based on temperature and maturation history. Petroleum generation stages were calculated assuming mainly type III kerogen and using a reaction kinetic data set based on Burnham [18]. The modeled hydrocarbon generation from Abu Madi-9 well is shown in Figure 11. This model shows that the corresponding to onset of the oil window (0.5-0.6 R_o) of the Miocene source rocks was during early Miocene at depths greater than 2000 m (Figure 11). The Miocene source rocks reached late mature stage at middle to upper Miocene age and the gas window in upper Miocene to lower Pliocene. Accordingly the hydrocarbon generation (oil and gas) started in middle to upper Miocene and peak hydrocarbon generation occurred during lower Pliocene (Figures 11 and 12).

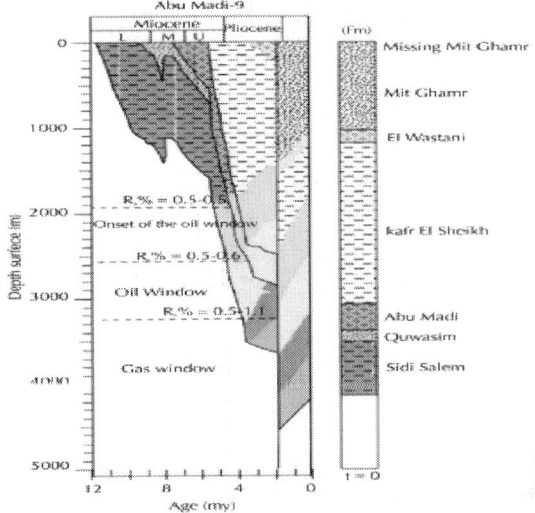

Figure 11: Burial history curve with hydrocarbon zones for the Abu Madi and Sidi Salem formations.

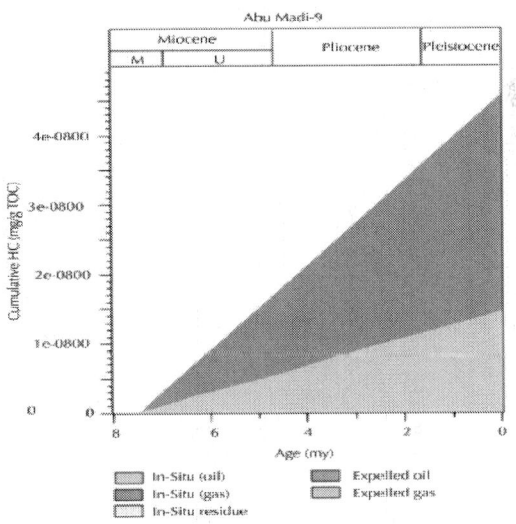

Figure 12: Calculated (cumulative model) of hydrocarbon generation from organic matter in the Abu Madi formation from Abu Madi-9 well, Abu Madi/ El Qar'a Gas field.

In the area of study geochemical analysis from two wells (Abu Madi-9 and Abu Madi-11) and burial history curves were taken to identify and characterize rich source intervals that probably the source of hydrocarbons, where organic matter (OM) in mudstone beds within Abu Madi Formation and shale beds of Sidi Salem formation is considered as effective hydrocarbon source rocks in study area.

MIGRATION AND ENTRAPMENT

Understanding the process of hydrocarbon generation and migration coupled with good geological data will assist to predict the ultimate hydrocarbon accumulations. The generation and migration of the hydrocarbons are thought to have reached their peak at the end of Miocene. This occurred after the main structural features had been imposed on the area and the main reservoirs had been deposited [12].

Migration along faults is responsible for vertical migration pathways from mature source rocks to shallower reservoirs (Figures 13 and 14). The presence of deep-cut channels, old valleys during Messinian, and the unconformities allows a good path for lateral updip migration and further entrapment in the shallow closures [12].

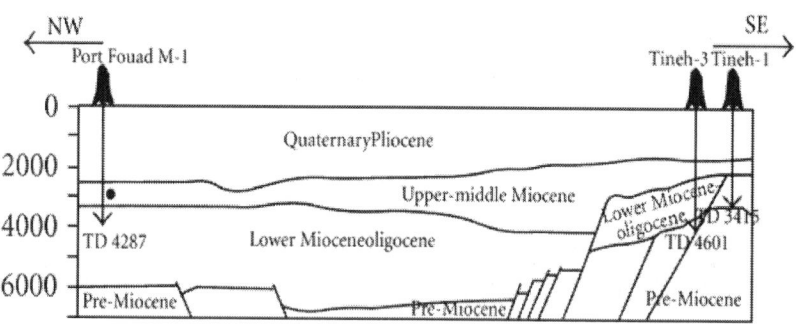

(a)

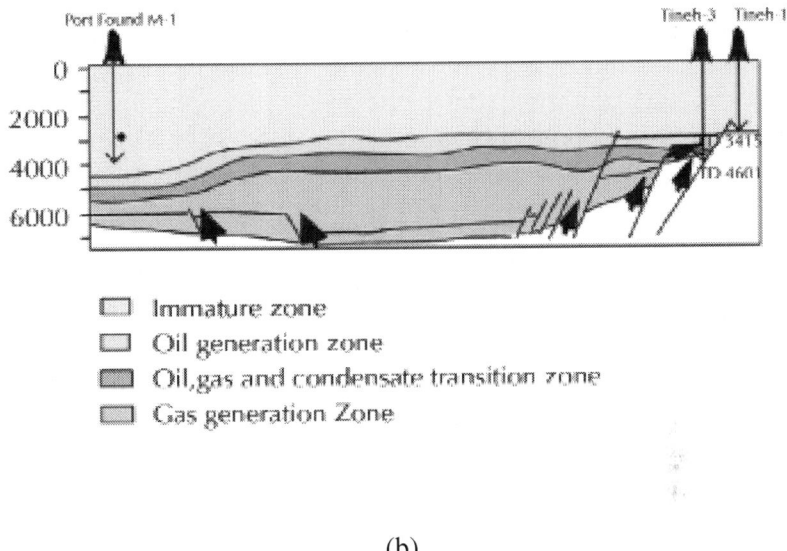

(b)

Figure 13: NW-SE cross sections showing the migration paths through the Nile Delta (modified after [12]).

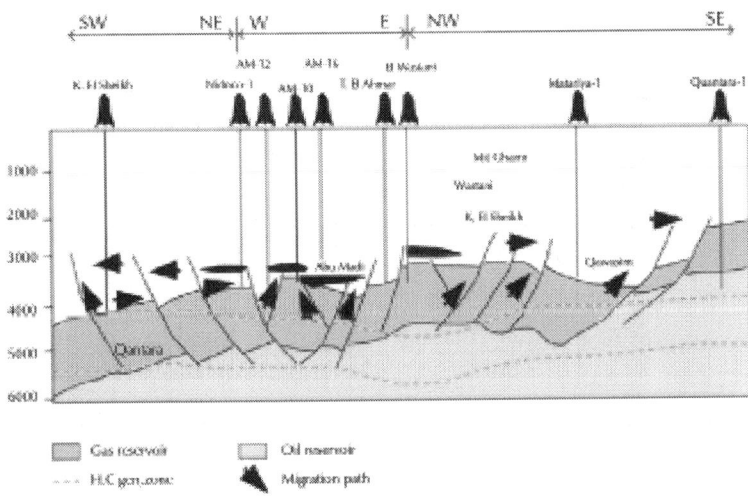

Figure 14: Hydrocarbon reservoirs and migration paths through the Nile Delta (modified after [12]).

In the study area, where the shallow Pliocene growth faults are not well developed, only the normal fault pattern plays the main role in the vertical migration for the generated hydrocarbons and its ultimate entrapment and accumulation in the Miocene and older reservoirs. The possibilities of finding out commercial Pliocene gas accumulation are relatively low and depending on the presence of syndepositional growth faults in the Pliocene.

CONCLUSIONS

Majority of Abu Madi kerogen belongs to mature type III in the principal zone of oil generation (oil window), R_o values range from 0.5 to 1%. With small amounts of kerogen immature type III, where R_o is less than 0.5%. Plot of hydrogen index versus oxygen index also indicates the predominance of organic matter of type III.

Subsidence history of Nile Delta basin classified into two phases, the first phase from early Cretaceous to middle Miocene is a mechanical (faulted-controlled) subsidence phase prevailed which continued from subsidence initiation (9 Ma) to (3 Ma). The second phase is a nonmechanical (thermal-controlled) subsidence from (3 Ma) and continued through Tertiary.

Numerical modeling of various wells indicates that onset of the oil window (0.5-0.6 R_o) of the Miocene source rocks was during early Miocene at depths greater than 2000 m. The Miocene source rocks reached late mature stage at middle to upper Miocene age and the gas window in upper Miocene to lower Pliocene. Accordingly the hydrocarbon generation (oil and gas) started in middle to upper Miocene and peak hydrocarbon generation occurred during lower Pliocene.

The maturation modeling of the study area revealed that the hydrocarbon compositions (gas and condensate) of Abu Madi formation are sourced from both mudstone beds in Abu Madi Formation and shale beds of Sidi Salem formation, where vitrinite reflectance estimated from models Maturity degree of Abu Madi Formation is early mature according to Vitrinite Reflectance determined from models (R_o range from 0.5 to 0.76) and Sidi Salem formation to main gas generation zone (R_o range from 1.3 to 2.6).

REFERENCES

1. Z. M. Zaghloul, A. A. Taha, and A. Gheith, Microfacies studies and paleo-environmental trends on the subsurface sediments of Kafr el sheikh well no. 1, vol. 5 of Nile Delta Area, Bulletin, Faculty of Science, Mansoura University, 1977.
2. L. M. Sharaf, "Source rock evaluation and geochemistry of condensates and natural gases, offshore Nile Delta, Egypt," Journal of Petroleum Geology, vol. 26, no. 2, pp. 189–209, 2003.
3. I. El-Heiny and N. Enani, "Regional stratigraphic interpretation pattern of Neogene's sediments, Northern Nile Delta, Egypt," in Proceedings of the 13th EGPC, Exploration and Production Conference, Cairo, Egypt, 1996.
4. EGPC (Egyptian General Petroleum Cooperation), Nile Delta and North Sinai: A Field, Discoveries and Hydrocarbon Potentials (A Comprehensive Overview), Egyptian General Petroleum, Cairo, Egypt, 1994.
5. M. Alfy, F. Polo, and M. Shash, "The geology of Abu Madi Gas Field," in Proceedings of the 11th Petroleum Exploration and Production Conference, EPGC, Cairo, vol. 2, pp. 485–513, 1992.
6. M. Sarhan and K. Hemdan, "North Nile Delta structural setting and trapping mechanism, Egypt," inProceedings of the 12th Petroleum Conference of EGPC, vol. 12, no. 1, pp. 1–17, Cairo, Egypt, 1994.
7. K. A. Khaled, et al., "Maturation and source-rock evaluation of mudstone beds in Abu Madi and Kafr el Sheikh Formations," in Proceedings of the Abu Madi Gas Field, Nile Delta, Egypt, Annals of the Egyptian Geological Survey, vol. 26, pp. 449–474, 2004.
8. J. Espitalie, M. Madec, B. Tissot, J. J. Mennig, and P. Leplat, "Source rock characterization, methods of petroleum exploration," in Proceedings of the Offshore Technology Conference, vol. 3, no. 9, pp. 439–444, May 1977.
9. K. E. Peters and M. R. Cassa, "Applied source rock geochemistry," in The Petroleum System, from Source to Trap. AAPG Mem, B. Magoon Leslie and W. G. Dow, Eds., pp. 93–120, 1994.
10. J. W. Cowie and M. G. Basset, "Global stratigraphic chart with geochronometric and magneto-stratigraphic calibration," in International Union of Geological Sciences, vol. 12, no. 2, 1989.

11. M. A. Yukler and Welte, "A three—dimensional deterministic dynamic model to determined geologic history and hydrocarbon generationmigration and accumulation in a sedimentary basin: Fossil Fuels," 1980.
12. H. Kamel, T. Eita, and M. Sarhan, "Nile Delta hydrocarbon potentiality," in Proceeding of the 14th Petroleum Conference, vol. l2, pp. 485–503, EGPC, Cairo, Egypt, October 1998.
13. J. M. Hunt, Petroleum Geochemistry and Geology, W. H. Freeman, New York, NY, USA, 2nd edition, 1996.
14. J. Espitalie, "Use of Tmax as a maturation index for different types of organic matter-comparison with vitrinite reflectance," in Thermal modeling in Sedimentary Basins, J. Burrus, Ed., pp. 475–496, Editions Technip, Paris, France, 1985.
15. D. W. Van Krevelen, "Organic geochemistry-old and new," Organic Geochemistry, vol. 6, pp. 1–10, 1984.
16. B. P. Tissot and D. H. Welte, Petroleum Formation and Occurrence, Springer, 2nd edition, 1984.
17. F. I. Metwalli and J. D. Pigott, "Analysis of petroleum system criticals of the Matruh-Shushan Basin, Western Desert, Egypt," Petroleum Geoscience, vol. 11, no. 2, pp. 157–178, 2005.
18. A. K. Burnham, A simple kinetic model of petroleum formation and cracking Lawrence Livermore National Laboratory Report UCID-21665, 1989.

Research of Drainage Gas Recovery Technology in Gas Wells

Shuren Yang[1], Di Xu1, Lili Liu[1], Chao Duan[2], and Liqun Xiu[1]

[1]Department of Petroleum Engineering, Northeast Petroleum University, Daqing, China
[2]CNOOC (China) Limited Zhanjiang Branch, Zhanjiang, China

ABSTRACT

Drainage gas recovery technology is the main method of gas recovery process in gas field, which has domestically and internationally been the main gas recovery processing measure in later stage of gas field production. In this context, produced water or condensate liquid will not be carried out of pit shaft by natural gas with the gradual drop of gas reservoir pressure and natural gas flow velocity, thus they will remain in pit shaft and form the so-called "gas well gathered liquid". This fluid severely affects natural gas output and leads to the decline of oil field economic benefits, thus drainage gas recovery measure must be taken to increase gas well or even gas field output. It becomes the primary problem to be solved to select the best drainage gas recovery measure which can maximize gas field benefits and optimize gas well safety.

INTRODUCTION

The purpose of this research is to establish a set of gas drainage way to determine the optimum process technical solution methods and procedures, according to various special geological condition and the gas wells in the production status.

Building such a comprehensive software platform will make the methods and techniques of drainage gas recovery have a new progress, make the management work well established in the scientific basis, so as to reduce the unreasonable factors, improve the quality of the design of the gas well process, and improve the overall economic benefit of gas field. So, this topic research has a very important and realistic significance. The practice experience shows that "draining gas extraction technology" plays a large role for the gas well steady-yield, and improves the recovery. So how to choose dewatering gas technology of more in line with the gas well, stronger adaptability and larger displacement and good economic performance, has become an issue that is worth studying. The related literature and achievement is not much, so research on this topic has a very vital significance [1].

BASIC MATHEMATIC MODEL AND DESIGN PROCEDURE OF OPTIMIZED COLUMN DRAINAGE GAS RECOVERY PROCESS

From optimized tube drainage gas recovery theory we can know that critical flow rate, critical flow velocity, correlation flow rate and correlation flow velocity of gas well continuous drainage can be determined by formulas below [2]:

$$q_{kp} = 0.648 \left(\gamma_g ZT\right)^{-\frac{1}{2}} \left(10553 - 34158 \frac{\gamma_g P_{wf}}{ZT}\right)^{\frac{1}{4}} p_{wf}^{\frac{1}{2}} d_i^2 \tag{2-1}$$

$$u_{kp} = 0.03313 \left(10553 - 34158 \frac{\gamma_g P_{wf}}{ZT}\right)^{\frac{1}{4}} \left(\frac{\gamma_g P_{wf}}{ZT}\right)^{-\frac{1}{2}} \tag{2-2}$$

$$u_r = \frac{u}{u_{kp}} \tag{2-3}$$

$$q_r = \frac{q_{sc}}{q_{kp}} \tag{2-4}$$

When actual parameters of gas wells cannot reach critical flow parameters, rational oil tube diameter should be re-selected to ensure continuous drainage by formula below:

$$d_i = 1.2433(\gamma_g ZT)^{\frac{1}{4}} \left(10553 - 34158 \frac{\gamma_g P_{wf}}{ZT}\right)^{-\frac{1}{8}} \times P_{wf}^{-\frac{1}{4}} q_{sc}^{\frac{1}{2}}$$

(2-5)

Where: q_{sc}——gas volume flow rate under standard state, 10^3 m³/d;

q_{kp}——critical flow rate established when gas well continuous drainage under standard state, 10^3 m³/d;

q_r——dimensionless correlation flow rate of gas well;

u_{kp}——as volume flow rate of tubing shoe fracture at gas well bottom, 10^3 m³/d;

u——gas flow velocity under standard state of gas well, m/s;

u_r——dimensionless correlation flow velocity of gas flow at tubing shoe;

P_{wf}——abstract well bottom pressure at tubing shoe, MPa;

T, Z——gas abstract temperature (K) and deviation factor of tubing shoe at well bottom;

γ_g——natural gas relative density;

d_i——design oil tube inner diameter, cm.

We can design the continuous optimized tube Nomograph of outlet wells by from Formula (2-1) to formula (2-5). Applied design applied Formula (2-1) to Formula (2-5) and Nomograph is summarized below:

1. Based on flowing tube diameter d_i, well depth H_i, output q_{sc}, well bottom flowing pressure P_{wf}, natural gas relative density γ_g and other parameters. We can get gas well continuous drainage flow rate q_{kp} and correlation parameter q_r by Formula (2-1) and Nomograph, thus we can judge gas well working system and drainage capacity;

2. When $q_r \geq 1$ and gas well can lift liquid continuously, and gas well can achieve relative stable "tri-stable" working system of pressure, production and gas/water ratio. When $q_r < 1$, gas well can not lift liquid continuously, we could re-optimize tube diameter d_i by Formula (2-2) and Nomograph and repeat procedure (1) to ensure $q_r \geq 1$, thus regular production could go on with new flowing tube diameter d_i;

3. From the possible established maximum pressure drop ($\Delta P = P_{wf} - P_{wh}$) of gas well, checkout whether inlet pressure is higher

than transiting pressure at the flowing tube diameter we got to ensure that consumers or collecting lines can get natural gas. If well head pressure is larger than transiting pressure, then the calculated results can be adopted, or gas well will select lager diameter by Formula (2)-(5);

For large water production gas wells, when it can not work regularly even large diameter oil tubes are applied, we can work out q_r by gas well equivalent oil tube diameter. When $q_r \geq 1$, and there is no danger to erode casing tubes, casing tubes can be used in production. With technical clues above, work out a computer program to design continuous drainage column [3].

THE MODEL ESTABLISHMENT FOR CRITICAL FLOW CARRYING LIQUID IN GAS WELLS

In 1969, Turner compares the model of the constant moving liquid films in the walls and the model of gas current of high velocity carrying liquid. The liquid films model describes the process for films to climb up to wall, but the calculation is difficult. The model of gas current of high velocity carrying liquid describes the drops in the center of the high-velocity gas current. Both the model of the liquid films and the model of gas current of high velocity carrying liquid exist in the actual production. Besides, the films in the walls will exchange medium with the drops in the gas current. The films descend and break into drops. Plenty of researches indicate that the model of gas current of high velocity carrying liquid is better for the problem of liquid loading in the gas wells. Turner presumed the drops in the wells as balls and deduced the calculations of the minimum gas flow rate and the minimum production if the gas can carry the liquid in the gas wells. Later Professor Li Min in Southwest Petroleum University proposed the ellipsoid model while Engineer in Liaoning Oil Exploration Bureau the cone model.

The chapter discusses Turner's ball model, Professor Li Min's ellipsoid model and Engineer Zhongyi Wang's cone model and provides the formulas of forecasting the minimum flow rate and critical production.

The Force Analysis of Drops

The drops in the gas wells are under the stress of the buoyancy from ambient gas F_g, their own gravity G_w and the drag force of gas F_D. Where:

$$G_w = \rho_l V g \qquad (3\text{-}1)$$

$$F_g = \rho_g V g \qquad (3\text{-}2)$$

$$F_D = C_D S \Delta p \qquad (3\text{-}3)$$

$$\Delta p = \rho_g v_g^2 / 2 \qquad (3\text{-}4)$$

where: V——the volume of the liquid column, m³;4
S——the drops' projected area in the direction of m4ovement, m²;
C_D——drag coefficient;
ΔP——the flowing pressure impacting the drops, Pa;
V_g——the velocity of gas current, m/s.

If the drops in the gas wells can be carried out of the wellbore, the sum of the buoyancy and the drag force is more than the gravity of drops, namely

$$F_g + F_D \cos\theta > G_w \qquad (3\text{-}5)$$

The following formula can be obtained after calculating:

# Research of Drainage Gas Recovery Technology in Gas Wells

$$u_{cr} > \sqrt{\frac{2(\rho_l - \rho_g)Vg}{C_D S \rho_g \cos\theta}} \qquad (3\text{-}6)$$

where: θ——hole deviation angle;
u_{cr}——critical flow rate carrying liquid in gas wells, m/s.

The critical flow rate carrying liquid in gas wells u_{cr} can be described as:

$$u_{cr} = \sqrt{\frac{2(\rho_l - \rho_g)Vg}{C_D S \rho_g \cos\theta}} \qquad (3\text{-}7)$$

In the Formula (3-7), ρ_l, ρ_g, θ can be obtained in the regular production. So the critical flow rate carrying liquid is mainly decided by the shape of the drops and the drag coefficient.

The Shape of Drops

1. The ball model Turner deduced the critical flow rate carrying liquid in the high gas-water ratio and flow in 1965. He assumed the drops in the gas wells as balls.
   Calculate the V and S of the ball in the Formula (3-7).

$$V = \frac{\pi d_d^3}{6} \qquad (3\text{-}8)$$

where: d_d——the diameter of the drops in the shape of ball, m.

$$S = \frac{\pi d_d^2}{4} \qquad (3\text{-}9)$$

Substitute (3-8), (3-9) into (3-7), the following formula can be reached:

$$u_{cr} = \sqrt{\frac{4(\rho_l - \rho_g)d_d g}{3C_D \rho_g}}$$

(3-10)

It can be seen in the Formula (3-10) that the flow rate needed for gas to carry drops is proportional to the diameter of drops. If the gas current can carry the biggest drop to the earth, then the liquid in the hole won't get together in the bottom of the well. How to determine the minimum diameter of the drops can be solved by Weber number. When the drops are carried upwards, they suffer two kinds of force. One is velocity pressure to crush the droplet, namely the inertia force, the other one is surface tension to keep it complete. Weber number is the ratio of these two forces.

$$N_{we} = \frac{v_g^2 \rho_g}{\sigma/d_d} = \frac{v_g^2 \rho_g d_d}{\sigma}$$

(3-11)

where: σ——the liquid surface tension, N/m.

After many experimental analyses, the researchers determined the critical value to maintain the drops as 30, namely the critical value of Weber number. It can be seen in the Formula (3-11) that the square of the gas flow rate is proportional to Weber number. When the gas flow rate is large enough to reach the critical Weber number, the drops will break under the inertia force. Substitute Weber number = 30 into (3-11) and work out the diameter d_d, which is the maximum diameter to maintain the drops steady.

$$d_{max} = \frac{30\sigma}{\rho_g v_g^2}$$

(3-12)

where: d_{max}——the maximum diameter of the ball drops, m.

Substitute (3-12) into (3-10), and deduce the minimum gas flow rate carrying the biggest drops.

Research of Drainage Gas Recovery Technology in Gas Wells

$$u_{ct} = \left[\frac{4g\sigma(\rho_l - \rho_g)}{C_D \rho_g^2}\right]^{0.25}$$

(3-13)

2. The ellipsoid model Li Min considers that when the drops move in the high velocity gas current, there are differential pressures before and after them. Liquid is flat under the pressure. The flat drop maintains under the surface tension and differential pressure. The equilibrium condition is as follows (The work acted on the drops for differential pressure is equal to the surface work as the surface tension of drops changes):

$$\Delta p S d h_d + \sigma dS = 0$$

(3-14)

where: h_d——the height of the flat drops, m.

As the drops become flat from ball and the volume is unchangeable, then

$$V = S h_d$$

(3-15)

According to the Formula (3-14), the following formula can be obtained:

$$\Delta p S/\sigma = -dS/dh_d$$

(3-16)

The following formula can be obtained from (3-15):

$$S = V/h_d$$

(3-17)

Derivate the two sides of (3-17) for h_d:

$$dS/dh_d = -V/h_d^2 = -S/h_d$$

(3-18)

From (3-16) and (3-18), the following formula can be reached:

$$h_d = 2\sigma/(\rho_g u_g^2) \quad (3\text{-}19)$$

Substitute $V/S = h_d = 2\sigma/(\rho_g u_g^2)$ into (3-7), the following formula can be obtained:

$$u_{cr} = \left[\frac{4g\sigma(\rho_l - \rho_g)}{C_D \rho_g^2}\right]^{0.25} \quad (3\text{-}20)$$

3. The cone model Zhongyi Wang model considers the deformation of drops movement, which is the same as Li Min model. However, Zhongyi Wang considers the drops as cones and calculates.

The following formula can be obtained after calculating:

$$V/S = 2\sigma/(3\rho_g u_g^2) \quad (3\text{-}21)$$

Substitute into (3-7):

$$u_{cr} = \left[\frac{4g\sigma(\rho_l - \rho_g)}{3C_D \rho_g^2}\right]^{0.25} \quad (3\text{-}22)$$

Theory Comparison of Three Continuous Liquid Carrying Models

Based on assumed conditions before and liquid force balance model, Professor Zhongyi Wang and Ming Li respectively build ball cap model and ellipsoid model. They actually are the selection of drag force factor C_D, which is mainly affected by turbulent flow, gas phase, compressibility and non-spherical grain. Comparison of the three models is shown in Table 1.

We can see from Table 1, the improvements of three models focus on the shape of liquid. About drag force factor, sphere model has the smallest drag force factor, while ball cap model has the largest. We can know that drag force factor increases dramatically with the raise of effective flow-facing areas.

Calculation of Critical Liquid Carrying Flow Rate

When daily output of a gas well is lower than critical liquid carrying flow rate, the gas well will gather liquid; the calculating results and practical situations comparison of three wells in three models are shown in Table 2.

From

Table 2 we can know that Turner model is most suitable for practical situations. The maximum critical liquid carrying flow rates and gathering-liquid judgments got from Ming Li model and Zhongyi Wang model have some deviation.

Critical liquid carrying flow velocity formula:

$$V_{cr} = 5.5 \left[\frac{\sigma(\rho_1 - \rho_2)}{\rho_g^2} \right]^{0.25}$$

(3-23)

Corresponding critical liquid carrying flow rate formula:

$$q_{cr} = 2.5 \times 10^4 \frac{A p V_g}{ZT}$$

(3-24)

Table 1: Continuous liquid carrying models comparison

Model	Liquid shape	Drag force factor	Critical liquid carrying velocity

Turner model	Sphere	0.44	$V_{cr} = 5.5 \left[\dfrac{\sigma(\rho_1 - \rho_2)}{\rho_g^2} \right]^{0.25}$
Ming Li model	Ellipse	1	$V_{cr} = 2.5 \left[\dfrac{\sigma(\rho_1 - \rho_2)}{\rho_g^2} \right]^{0.25}$
Zhongyi Wang model	Ball cap model	1.17	$V_{cr} = 1.8 \left[\dfrac{\sigma(\rho_1 - \rho_2)}{\rho_g^2} \right]^{0.25}$

Table 2: Comparison of calculating results and practical situations

Number	Daily output (10⁴m³/d)	Critical liquid carrying flow rate (10⁴ m3/d)						Practical results
		Zhongyi Wang model		Ming Li model		Turner model		
1	4.51332	1.4638	No liquid	2.0334	No liquid	4.4733	No liquid	No liquid
2	1.0445	1.4664	Liquid	2.0365	Liquid	1.4805	Liquid	Liquid
3	5.0904	1.95	No liquid	2.7362	No liquid	6.0196	Liquid	Liquid

where: q_{cr}——the minimum flow rate or unloading flow rate required for gas carrying liquid, $10^4 m^3/d$;

A——section area of oil pipes, m^2;

p——bottom hole flowing pressure, MPa;

T——bottom hole gas temperature, K;

Z——gas compressibility factor under specific bottom hole flowing pressure and gas temperature.

Formula (3-10) and (3-11) are minimum gas flow velocity and minimum gas flow rate practical formulas derived from thoughts of Turner and others. They are suitable for gas-water wells and gas-condensate oil wells.

Because of lacking physical property data of surface tension under different pressure and temperature, approximate calculation could adopt values below:

For water: $\sigma = 60 \times 10^{-3}$ N/m, $\rho_l = 1074$ kg/m³;

For condensate oil: σ = 20×10⁻³ N/m, $ρ_l$ = 721 kg/m³.

As the surface tension and density difference between gas and water are higher than that between oil and gas, physical values of well water mixed liquid are calculated by water for well which products both water and condensate oil.

For better applied in field, two simplified formulas are derived by Turner and others under several assumed conditions, the preconditions of simplified formulas are [4]:

1) $γ_g$ = 0.6, T = 120 °F, $ρ_g$ = 0.0031 P 2) For water: σ = 60 dyn/cm, $ρ_L$ = 67 lb/ft³

3) For condensate oil: σ = 20 dyn/cm, $ρ_L$ = 47 lb/ft³

Then, for water:

$$V_g = \frac{5.62 \times (67 - 0.0031p)^{0.25}}{(0.0031p)^{0.5}}$$

(3-25)

For condensate oil:

$$V_g = \frac{4.02 \times (45 - 0.0031p)^{0.25}}{(0.0031p)^{0.5}}$$

(3-26)

Minimum flow rate formula is:

$$q_{sc} = 3.06 \frac{A p V_g}{ZT}$$

(3-27)

where: q_{sc}——the minimum flow rate or unloading flow rate required for gas carrying liquid, 10⁶m³/d;

A——section area of oil pipes, ft²;

p——bottom hole flowing pressure, psi;

T——bottom hole gas temperature, °F;

Z——gas compressibility factor under specific bottom hole flowing pressure and gas temperature;

v_g——minimum unloading flow rate, ft/s.

OPTIMIZATION SOFTWARE OF DRAINAGE GAS RECOVERY

This software applicants Visual Basic 6.0 program, convenient operation and clear operator interface, somewhat similar to the Windows operating interface. The main interface is shown inFigure 1. The main menu of the software including "Introduce of Drainage Gas Recovery", "Computational Analysis", "Help" and "Exit" four options, in which "Computational Analysis" is used for calculations mentioned in the previous chapters (including Down-hole Pressure, Critical Flow, Liquid Loading and Column Optimizing). And the "Help" gives various considerations and other explanatory to software installation and operation. The "Exit" is used to end the software, and then return to Windows operating system [5].

Click "Computational Analysis" in the menu interface will appear an interface as follows.

Introduce each sub-menu's functions as follow:

1. Pressure-drop Calculation of Water-producing Gas Well This Program can calculate the bottom-hole pressure of the gas well using the input parameters. Show in Figure 2.
2. Calculation of Critical Delivery This Program can calculate the critical delivery and critical velocity of the gas well using the input parameters. Show in Figure 3.
3. Calculation of Liquid Loading This Program can calculate the liquid loading of the gas well using the input parameters. Show in Figure 4.
4. Column Optimizing This Program can calculate the liquid loading of the gas well using the input parameters. Show in Figure 5.

CONCLUSIONS

- According to the actual situation of gas field, the paper analyzes the wellhead back pressure effect on the critical liquid carrying

capacity, and Calculation formula of Critical carrying fluid flow velocity and critical fluid flow is derived;
- Using Turner model to establish critical liquid carrying flow velocity calculation formula of gas well combining field situations based on the character of gas well liquid carrying. We can also establish critical liquid

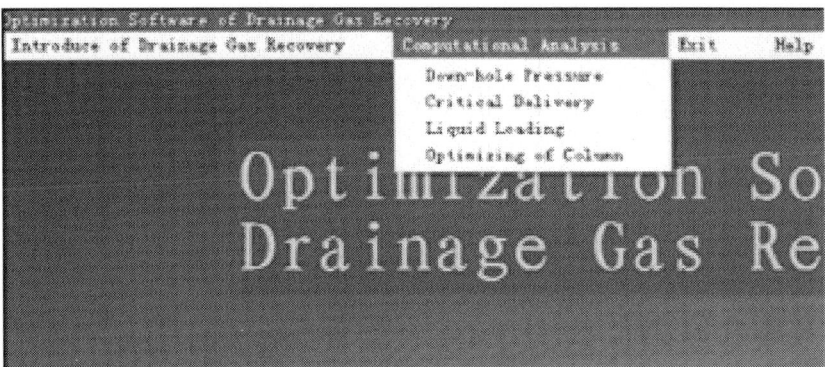

Figure 1: Calculation analysis.

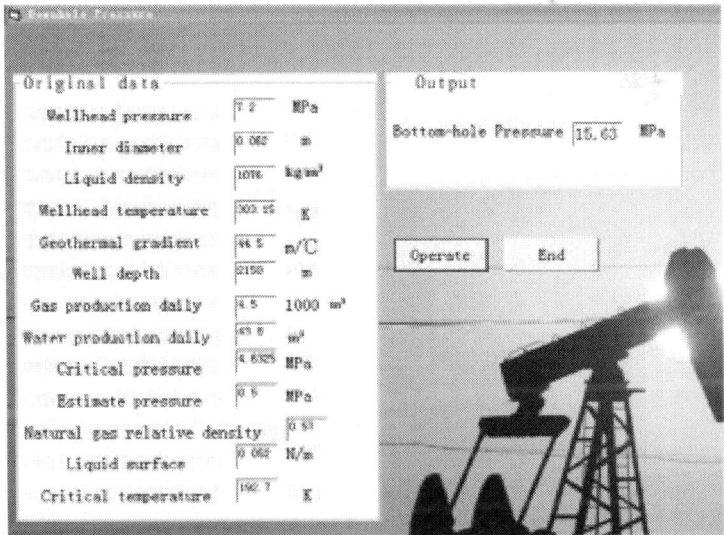

Figure 2: Interface of down-hole pressure calculation.

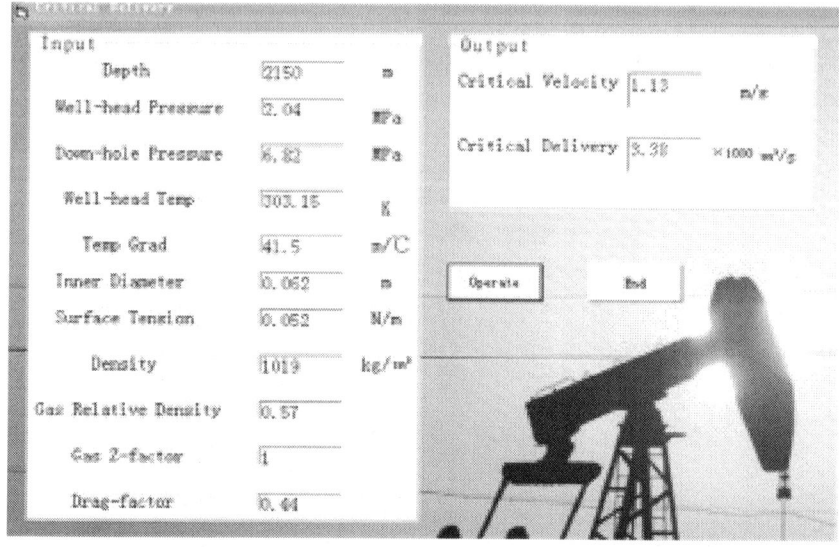

Figure 3: Interface of critical delivery calculation.

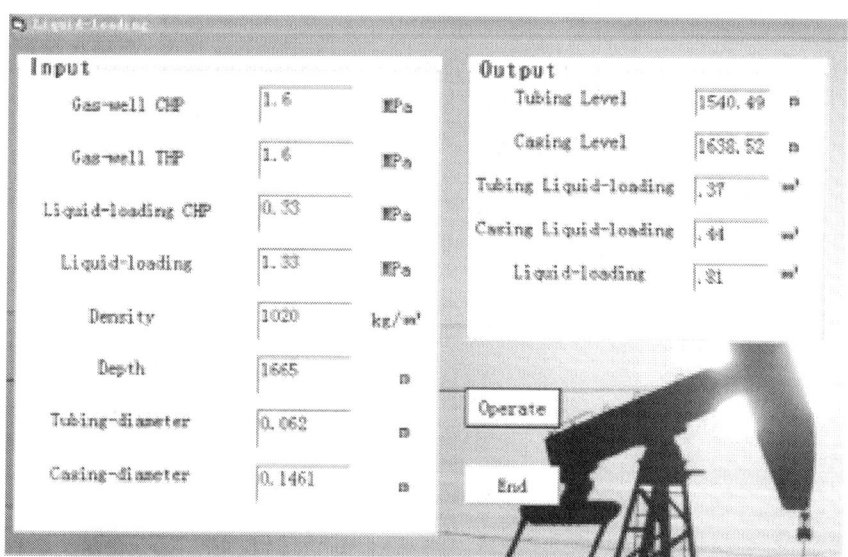

Figure 4: Interface of liquid loading calculation.

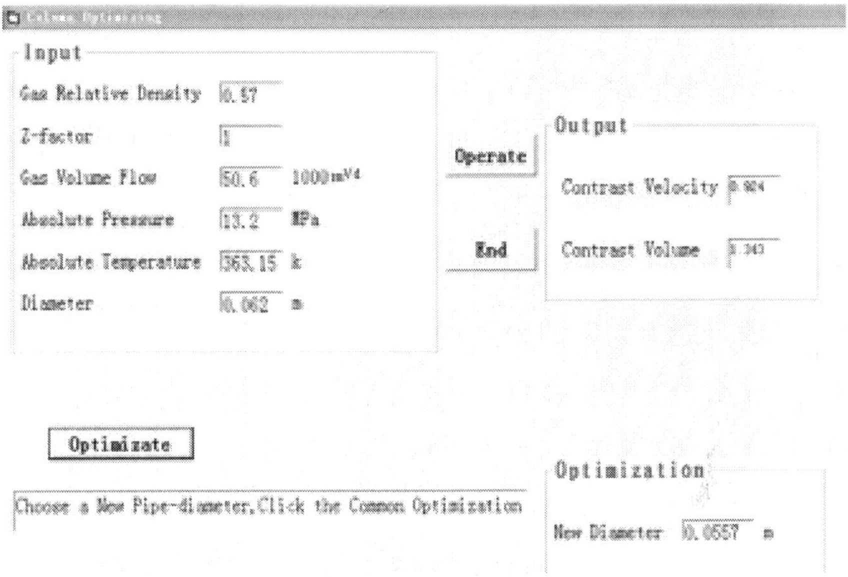

Figure 5: Interface of column optimizing.

Carrying flow rate calculation formula of gas well referring to critical liquid carrying flow velocity calculation formula of gas well. Practical application shows that established formulas could meet requirements of gas well liquid gathering judgment;

3. According to the models, the optimization designing optimized software of dewatering gas recovery suits for the majority of oil and gas field.

REFERENCES

1. Lage, A.C.V.M. and Time, R.W. Mechanistic Model for Upward Two-Phase Flow in Annulus. SPE 63127.
2. Li, Z.Y., Ma, J.Z. and Liao, Z.C. (2008) Research of Drainage Gas Recovery Technology. Inner Mongolia Petrochemical Industry, 11.
3. Zhou, M.Q., Duan, Y.-G., et al. (2010) Research and Application on Drainage Gas Recovery by Gas Well Self-Energy. Journal of Chongqing Technology Institute (Natural Science Edition), 10.

4. Wu, Y.T. (2009) The Preference of Drainage Gas Recovery Methods and Safety Assessment of Gas Well with Water Drainage. China University of Petroleum, 4.
5. Zheng, X.X. (2008) Drainage Gas Recovery Method Optimization. China University of Petroleum, Beijing.

Alternative System of Industrial Paint Applied to Spherical Mount for Liquefied Petroleum Gas

Fernando B. Mainier[1], Francisco Otavio Pereira da Silva[2], Gilberto Oliveira da Silva[2]

[1]Escola de Engenharia, Universidade Federal Fluminense (UFF), Niterói, Brazil
[2]Petrobras—Petróleo Brasileiro SA, Rio De Janeiro, Brazil

ABSTRACT

The present article reports the application of zinc ethyl silicate paint and the use of internal and external paint schemes on carbon steel spheres for the storage of liquefied petroleum gas. The new paint scheme eliminates the steps of blasting in the field and minimizes the collection of waste generated and the environmental impact, reducing the service time onsite and therefore providing a productivity gain and better health and cleanliness at work. The results were obtained through test runs and qualified in bodies-of-proof made with the same characteristics as the sphere, that is, using the same material (carbon steel), thickness, and mechanical formation and subject to the same conditions of design and implementation process. The paint scheme was approved, qualified, and committed to the supplier's warranty with the paint manufacturer and assembler of the storage spheres for liquefied petroleum gas.

INTRODUCTION

Carbon steel has been the most widely used material in most segments of basic production assets of the society. And, in recent decades, there has been considerable progress in both the manufacture of new alloys and nonferrous alloys and the development of new composite materials. However, given the scope of the use of common carbon steel, it is expected that the field of exposure to deterioration also occurs widely. In the case of petroleum refineries and petrochemical plants, the study of the corrosion processes has a bigger place, when one takes into account that about 50% of the failures of materials are credited to corrosion. The process of applying knowledge to corrosion principles and anti-corrosion protection as well as rules about practical suitability has been a challenge in the field of engineering equipment [1] [2].

Carbon steel is the main material used in the manufacture of equipment and industrial pipes; however due to corrosion the possibility of industrial use is dependent on the use of anti-corrosion coatings, and industrial painting stands out among the anti-corrosion processes. Protective coatings are generally applied on metallic surfaces to form a barrier between the surface and the corrosive medium and therefore prevent or minimize the corrosion process [3] [4].

The coatings can be metallic, organic, inorganic, or composite and their use for corrosion prevention will depend on a number of factors such as the nature of the corrosive medium, temperature, pressure, material hardness, mechanical strength, thermal conductivity, electrical conductivity, cost, and so on.

Industrial painting can be defined as any composition of chemicals, both organic and inorganic, applied as a liquid or paste to form a film on the surfaces of materials, which will undergo subsequent hardening, forming a solid adherent coating that is able to protect the materials against various corrosive media. The thicknesses of the coatings on metallic surfaces can vary from 60 to 500 μm, depending on the use and the aggressiveness of corrosive media [5]

Industrial painting of field industrial equipment can be carried out by applying industrial coatings using mobile facilities that comprise abrasive blasting machines, manual or automatic spray guns, and other equipment necessary for the application of paints.

Industrial painting must be based on the principles of quality and premises related to standards, procedures, occupational health, industrial safety, and the environment. Therefore, industrial painting must be suited to the organizational process under which all steps of the processes adopted are planned, implemented, monitored, recorded, reported, and archived.

It is important to note that even with paint application standards, periodic inspection is essential in the monitoring of their performance against the corrosive medium conditions and estimated life cycle. On-the-spot inspection aims to assess the failures by corrosion as well as mechanical damage generated by the transport and other operations. The inspection must be carried out to the full extent of the application; however, special attention should be given to sharp corners, welded areas, cracks, edges, and so on. This article aims to show the advantages of painting metallic parts at the plant and apply the final welding on the field.

MANUFACTURE AND ASSEMBLY OF THE SPHERES CONSIDERING THE USE OF THE CONVENTIONAL SCHEME PAINT AND THE SHOP PRIMER SCHEME PAINT

The painting of spheres for storage of liquefied petroleum gases is usually carried out on the construction site. The lining of this equipment is based on the operating conditions, environmental conditions, and costs, and its scope includes the treatment of surfaces, the paint, and its application.

Petrobras Standard N-1375 [6] defines the paint schemes of the liquefied gas storage spheres as environmental and operational conditions. In the case of base paint (shop primer), liquid paint (ethyl silicate inorganic zinc) is used on the basis of the Petrobras Standard N-1841 [7]

This paint has a high content of metallic zinc in the dry film of zinc-rich coatings (minimum 85% Zn by weight), which provides a greater weather resistance; however, it interferes directly in the operation of oxy-cutting (cutting of sheets and pipes using a blowtorch with oxidizing gas mixtures) and welding of the plates. In the process of oxy-cutting, torch nozzle clogging can occur, while at low speeds the formation of pores can take place, forcing the removal of welded joints.

In addition, due to the high temperatures of these processes the formation of toxic fumes (zinc in gaseous form) occurs, damaging the health of workers involved directly with the welding process. Classification societies are non-governmental organizations that establish and maintain technical standards for the construction and operation of ships, offshore structures, petroleum refineries, and so on. These classification societies do not certify this paint for use in welding operations, and therefore there can be a risk to the health of workers.

The proposed process uses a modified paint formulation with low zinc content which has excellent anti-corrosion properties and is compatible with oxy-cutting processes and automatic welding [8]. Due to the low formation of toxic fumes (determined by laboratory analysis), international classifications societies have awarded the welding certificate.

Conventional Scheme without Application of Painting "Shop Primer"

In the conventional system, the storage spheres of liquefied petroleum gas are assembled from modules of carbon steel sheets without application of the paint system, as shown in Figure 1 and Figure 2.

After industrial assembly of the modules, heat-treatment of weld beads is carried out and then the entire sphere is blasted with steel shot. Subsequently, the base paint (shop primer) and then finally the finishing paint are applied.

Proposed Scheme of Assembly with the Application Modules Painted with Ethyl Zinc Silicate Paint (Modified Shop Primer)

The new proposed process consists, essentially, of two phases. In the first phase, the carbon steel modules are blasted with steel shot; then they are painted at the factory (Figure 3) and transported for the assembly into a sphere in the field (Figure 4).

Figure 1: Industrial assembly of the sphere modules without paint application.

Figure 2: Sphere modules without paint application.

Figure 3: Painting of the module at the factory.

Figure 4: Assembly of the ready-painted module on the field.

The second phase in the field, the assembly sequence, consists of the following steps: welding of modules, heat treatment in weld beads, painting of weld beads, hydro blasting with low-pressure water over the applied paint, and finally finish painting (Figure 5 and Figure 6).

QUALIFICATION OF THE SPHERES ASSEMBLY PROCEDURES USING PROOF-BODIES PAINTED WITH ZINC SILICATE ETHYL MODIFIED PAINT

The methodology of qualification procedures for painting of sphere assembly modules at the factory essentially consists of the preparation of bodies-of-proof (BPs) with the same carbon steel and same thicknesses following the same procedures as were carried out in the assembly of the sphere in the field. Such processes are based on Procedure CQEQ-

064 Petrobras (Qualification of Procedure of Storage Sphere Painting of Liquefied Petroleum Products [9].

The BPs used in the experiments was removed from a surplus (unused) part of the sphere assembly after the confirmation process and cut by oxy-fuel cutting into a sample with dimensions of 1400 × 800 × 50 mm, as shown in Figure 7. Then, the BP was blasted with steel shot based on Standard ISO 8501 [10], thus forming a roughness profile in the range of 40 to 70 μm.

In the preparation of BP it is essential to evaluate the local environmental conditions (relative humidity, temperature, dew point, and temperature of carbon steel sheet) prior to application of the paint, using as reference the Petrobras Standard N-0013 [11]

Figure 5: Application of finishing paint

Alternative System of Industrial Paint Applied to Spherical Mount for...

Figure 6: Final finished painted sphere.

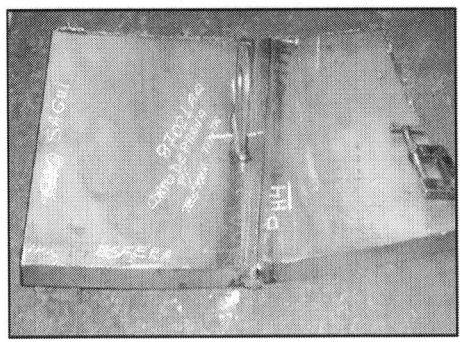

Figure 7: The body-of-proof (BP) used in the experiments (qualification).

After surface preparation, one 25-to-30-μm coat of ethyl silicate zinc modified paint (shop primer paint) was applied using a spray gun with a tank mechanical stirrer and spray pressure of 40 psi (Figure 8).

Back to reinforce the procedures adopted in the body-of-proof (BP) are identical to those adopted for the sphere assembly.

Besides the field and assembly simulation of all procedures performed on the sphere, the body-of-proof (BP) was cut in half, bevelled, and welded with the same characteristics as those used on the sphere project construction site.

To further verify the behaviour of paint in real assembly conditions, the painted BP was submitted to heat treatment (Figure 9) in the same conditions of heat treatment as for the sphere assembly.

One of the most important points in this work is to qualitatively and quantitatively verify the paint behaviour considering that the soldering temperature and subsequent heat treatment can reach temperatures of 650°C. The biggest expectation would know how "shop primer paint" would behave after heat treatment, reaching temperatures of approximately 650°C. However, one of the features of this paint is the ability to withstand high temperatures.

As shown in Figure 10, after removal from heat treatment, the body-of-proof (BP) presented dark spots; however, adhesion tests carried out on the basis of the standards ABNT NBR 11003 [12] and Petrobras N-0013 [13] showed good results.

Painting procedures after heat treatment and application of one coat of epoxy-zinc phosphate paint as the base consisted of the following steps. The first was cleaning of the surface of the BP by hydro blasting with fresh water (pH between 6.5 and 7.5) at a pressure of 3000 to 4000 psi. After full drying one coat of epoxy-zinc

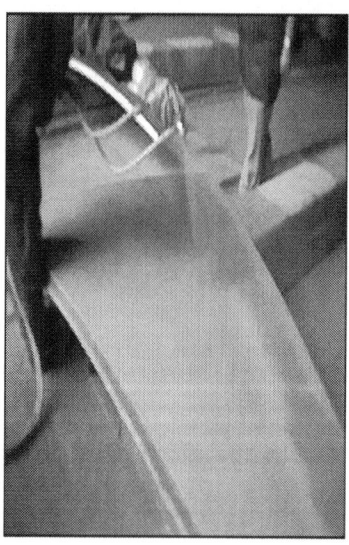

Figure 8: Body-of-proof (BP) painting with spray gun.

Figure 9: Body-of-proof (BP) under heat treatment.

Phosphate paint was applied to give a film with a minimum thickness of 100 μm over the outside of the BP.

The application method used a conventional pistol with a mechanical stirrer and environmental conditions occurred based on the Petrobras standard N-0013 [11] and ABNT NBR 11003 [12].

The purpose of this test is to measure the mechanical tensile strength of a coating. The sample is subjected to increasing tensile stresses until the weakest path through the material fractures. The acceptance criterion for pull-off adhesion testing using the ASTM D-4541 [13] is that a value of at least 12 MPa must be achieved. The results obtained for 13.6, 15.6, and 15.1 MPa are shown in Figure 11.

EVALUATION OF ASSEMBLY PROCESS WITH THE MODULES PAINTED WITH ETHYL SILICATE ZINC MODIFIED PAINT (SHOP PRIMER PAINT)

Through the application of the methodology presented with the qualification of the procedure, the paint scheme with modified shop primer paint has been implemented in the manufacture and assembly of liquefied petroleum gas storage spheres in oil and refinery.

The results were considered excellent compared to conventional scheme paint, providing the following improvements:

- elimination of blasting activity in the field;
- reduction in labour costs by approximately 2400 person hours;
- reduction of the amount of painting work done in a confined space;

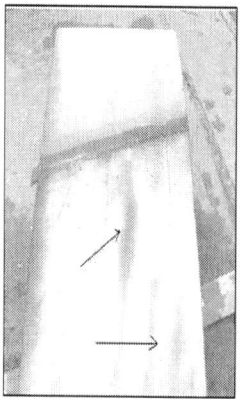

Figure 10: The BP presented dark spots after heat treatment.

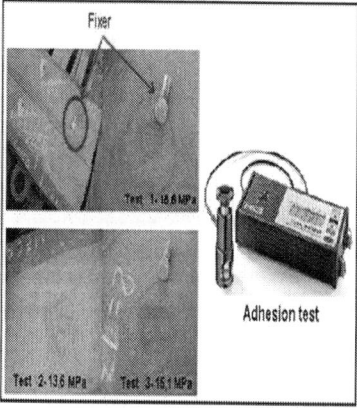

Figure 11: Following the adhesion test.

- reduction of noise pollution and noise index;
- reduction of polluting agents in the field, where they are more difficult to control (whereas in the new scheme most of the modules are welded at the factory, where the highest level of control is possible);
- equality of the roughness profile for the application of paint because this was obtained in the factory under better technical and operational conditions than on the field;
- reduction of delivery time by at least 30 days.

CONCLUSIONS

Due to the need for refineries to expand the production of petroleum gas storage spheres, in a short time, the manufacture and assembly of storage spheres has become one of the main objectives of equipment manufacturers.

Through this challenge and based on the excellent test results, an alternative system that achieves this goal has been implemented, improving work safety and leading to higher productivity and better quality of the environment. These factors have resulted in clean conditions, providing a considerable reduction in unsafe conditions and waste generation.

REFERENCES

1. Terzi, R. and Mainier, F.B. (2008) Internal Corrosion Monitoring Offshore Platforms. Tecno-Lógica, Santa Cruz do Sul, 14-21 (in Portuguese).
2. Roberge, P.R. (2000) Handbook of Corrosion Engineering. Vol. 1128. McGraw-Hill, New York.
3. Gentil, V. (2011) Corrosion. 6th Edition, LTC Livros Técnicos e Científicos (Publisher), Rio de Janeiro (in Portuguese).
4. Mansfeld, F. (2003) Electrochemical Methods of Corrosion Testing. ASM Handbook, 13, 446-462.
5. Talbert, R. (2007) Paint Technology Handbook. CRC Press, Boca Raton.http://dx.doi.org/10.1201/9781420017786

6. Petrobras Standard N-1375 (2007) Painting of Sphere and Cylinder for Liquefied Gas Storage Derived from Oil and Ammonia. Technical Standardisation Committee of Petrobras (in Portuguese).
7. Petrobras Standard N-1841 (2007) Shop Ethyl-Silicate Zinc Primer. Technical Standardisation Committee of Petrobras (in Portuguese).
8. Sadler, H. (2007) Sorting out Certifications for Welding Consumables. Welding Journal, 86, 42-45.
9. Petrobras Standard CQEQ-064 (2007) Qualification of Procedure of Storage Sphere Painting of Liquefied Petroleum. Technical Standardisation Committee of Petrobras (in Portuguese).
10. ISO 8501 (2000) Preparation of Steel Substrates before Application of Paints and Related Products.
11. Petrobras Standard N-0013 G (2004) Technical Requirements for Painting Services, CONTEC, Technical Standardisation Committee of Petrobras (in Portuguese).
12. ABNT NBR 11003 (1990) Determination of adherence. Brazilian Association of Technical Standards (in Portuguese).
13. ASTM D 4541 (2000) Standard Test Method for Pull-Off Strength of Coatings Using Portable Adhesion Testers.

Chapter 4

Life-Cycle Analyses of Energy Consumption and GHG Emissions of Natural Gas-Based Alternative Vehicle Fuels in China

Xunmin Ou[1,2] and Xiliang Zhang[1,2]

[1]Institute of Energy, Environment and Economy (3E), Tsinghua University, Beijing 100084, China
[2]China Automotive Energy Research Center, Tsinghua University, Beijing 100084, China

ABSTRACT

Tsinghua life-cycle analysis model (TLCAM) has been used to examine the primary fossil energy consumption and greenhouse gas (GHG) emissions for natural gas- (NG-) based alternative vehicle fuels in China. The results show that (1) compress NG- and liquid NG-powered vehicles have similar well-to-wheels (WTW) fossil energy uses to conventional gasoline- and diesel-fueled vehicles, but differences emerge with the distance of NG transportation. Additionally, thanks to NG having a lower carbon content than petroleum, CNG- and LNG-powered vehicles emit 10–20% and 5–10% less GHGs than gasoline- and diesel-fueled vehicles, respectively; (2) gas-to-liquid- (GTL-) powered vehicles involve approximately 50% more WTW fossil energy uses than conventional gasoline- and diesel-fueled vehicles, primarily because of the low efficiency of GTL production. Nevertheless, since NG has a lower carbon content than petroleum, GTL-powered vehicles emit approximately 30% more GHGs than conventional-fuel vehicles; (3) The carbon emission intensity of the LNG energy chain is highly sensitive to the efficiency of NG liquefaction and the form of energy used in that process.

BACKGROUND

Alternative Vehicle Fuels in China

General Background

During the period of the eleventh five-year (2006–2010) plan, the Chinese automobile market experienced strong growth. Vehicle ownership increased 20% annually during that period, amounting to 87 million by the end of 2010. This strong growth promoted a steadily increasing demand for fuels and, accordingly, a substantial escalation in petroleum prices. Meanwhile, pollutant and carbon dioxide (CO_2) emissions associated with petroleum combustion also posed major environmental concerns. Consequently, alternative vehicle fuels are gaining increasing interest.

Currently, several nonconventional fuels have been marketed as alternatives to regular vehicle fuels (i.e., gasoline and diesel); these include vehicular natural gas—compressed natural gas (CNG) and liquefied natural gas (LNG)—biodiesel, methanol-gasoline blends, ethanol-gasoline blends, and coal-derived fuels. Electric vehicles have also been introduced as a potential solution to the disadvantages of vehicles consuming conventional fuels.

The study of CAERC (2012) [1] found that the ownership of natural gas vehicles in China exceeded 500,000 as of 2010; annually 4.6 million tons of vehicular gasoline and diesel are being replaced by natural gas in 2010.

Moreover, the annual consumption of methanol-gasoline blends, biofuels, and coal-derived fuels was 200 million tons, 1.83 million tons, and 33,000 tons, respectively. The annual saving of gasoline and diesel by electric vehicles is less than 40,000 tons. In total, these alternative fuels replace only a fraction (3.3%) of China's vehicular gasoline and diesel consumption. However, the acceptance of alternative vehicle fuels is profoundly affected by industrial productivity and governmental policies. Furthermore, their acceptance may have a substantial impact on local liquid fuel markets.

Current Status

Two types of natural gas products have received attention as alternative vehicle fuels—CNG and LNG.

CNG Vehicle Fuel. According to the NGV Global (2012) [4] statistic, as of June of 2010, the ownership of natural gas vehicles in China exceeded 1,104,000; 98.9% of them are CNGV. Currently, CNG vehicles are in use in 30 provinces (cities or autonomous regions) of China. As of 2010, the officially registered CNG vehicle ownership amounted to 20,000. Taking unregistered vehicles into account (i.e., CNG vehicles converted from regular-fuel autos), the total number of CNG vehicles exceeded 500,000. Meanwhile, CNG station development has made considerable progress. As of 2009, there were 1,055 CNG stations across China, which was 500 more than there were at the end of 2007.

Statistical data of China LNG Vehicle Net (CLNGVN, 2012) [5] in 2010 from several regions where CNG vehicles have been relatively popular (i.e., Sichuan, Chongqing, Harbin, Urumqi, and Xi'an) show that CNG vehicles are primarily used as city buses (replacing diesel buses), taxicabs (replacing gasoline taxis), and governmental automobiles. Approximately 80% of these vehicles were used for commercial purposes.

According to the NGV Global (2012) [4], as of June of 2012, the ownership of natural gas vehicles in China exceeded 1.1 million; 98.9% of them are CNGV.

Current high gasoline prices reinforce the economic advantages of CNG vehicles. As a result, these vehicles have been well marketed in regions with abundant natural gas sources. However, owing to technical limitations (e.g., fuel availability and difficulties in fuel station construction), CNG vehicles are primarily suitable for city buses and short-distance transport. As of 2010, there were 600,000 CNG vehicles in China, which consumed 6 billion cubic meters of CNG annually as a replacement for 2.76 million tons of gasoline and 260,000 tons of diesel.

LNG Vehicle Fuel. LNG vehicles have not been widely utilized in China, primarily owing to the lack of LNG sources, the high cost of conversion to LNG, and lack of LNG fueling stations. According to a survey of seven Chinese cities in 2010, 2,800 LNG vehicles were in operation, of which 56% were city buses; a significant proportion of the remainder were accounted for by heavy trucks. For example, 400 LNG heavy trucks were used by the Guanghui Industry Investment Group (Xinjiang, China) for LNG delivery, and point-to-point transport between coal plants. In 2010, several cities (e.g., Erdos, Zhuhai) undertook pilot programs to promote the replacement of heavy diesel trucks by LNG alternatives.

Thus far, LNG vehicles are primarily being operated at the demonstration stage. Although LNG terminals began operation in Jiangsu and Dalian in 2011, they mainly targeted specific industrial users.

In recent 2 years, LNG vehicle market has developed very quickly and the ownership of LNG vehicles in China exceeded 70 thousand million; the numbers of LNG refueling stations reached to about 500 as the end of October of 2012 (CLNGVN, 2012) [5].

Consequently, LNG has played a more and more important role as an alternative to diesel as a fuel for vehicles.

Impact of Alternative Vehicle Fuels on Gasoline Demand in China

In 2010, the above-mentioned alternative fuels served as a replacement for approximately 3.3% (i.e., 7.1 million tons) of the total annual consumption of regular vehicle fuels. Specifically, gasoline alternatives (CNG, methanol, ethanol, and electricity) reduced annual vehicular gasoline consumption by approximately 6.6% (i.e., 60,000 tons). Diesel alternatives (LNG, coal-derived fuel, and biodiesel) were used to replace approximately 1.8% of annual vehicular diesel consumption. Of the various alternative vehicle fuels, CNG appears to be the most successful, accounting for 83% of gasoline substitution and 71% of that for diesel.

Life-Cycle Studies on Vehicle Fuels

Life-cycle analyses (LCAs) of energy consumption and greenhouse gas (GHG) emissions are an important component in assessing vehicle fuel pathways. Many studies have produced results that pertain to specific geographic locations (Zhang et al., 2008) [2].

In the last two decades, there have been intensive studies of alternative fuels and related vehicle technologies. Models have been used to analyze fuel use and carbon emissions of various alternative-fuel routes, such as the life-cycle emission model (LEM) and the greenhouse gas, regulated emissions, and energy use of transportation energy (GREET) model [6–9]. Many organizations and research groups have used the two models in their LCAs of alternative-fuel vehicles in various regions, such as Europe and North America [10, 11]. Their findings are highly region specific, and thus cannot be directly applied to other locations. By employing different models, researchers have investigated the LCA results of different energy pathways, including the natural gas pathway globally [12–21].

In China, earlier LCA studies focused on single-route analyses of passenger cars, new-energy vehicles, and vehicle operation (e.g., engine bench tests). Recent studies have increasingly focused on two- and multiple-route comparisons. However, these recent studies have lacked statistical information or basic data reflecting actual vehicle operations; many conclusions were drawn from experiments or predictions. Consequently, it is difficult to make a direct comparison between the findings from different studies.

Recently, Tsinghua University collaborated with Ford, General Motor, and the China Automotive Technology and Research Center (CATARC) to perform comparative LCAs for multifuel and vehicle pathways of vehicles using the GREET model. Because the original GREET model and default parameters were designed based on the energy production chain in the United States, the Chinese research team maximally incorporated localized data into its analyses. Thus, the results generally reflect the actual operation conditions in China [2, 3, 22].

TSINGHUA LIFE-CYCLE ANALYSIS MODEL AND KEY EMISSION INTENSITIES

The present study investigated the energy use and GHG emissions of various vehicle fuel pathways using the well-to-wheels (WTW) method and related LCA tools. This study focused only on LCA and included no subsequent evaluations.

The group has developed an LCA model termed the Tsinghua life-cycle analysis model (TLCAM), which is specifically designed to analyze vehicle fuels in China. The model is analogous to the GREET model (Wang, 1999)—a transportation energy model specific to the United States—and is implemented using Microsoft Excel (Microsoft, Redmond, WA, USA). TLCAM has been applied to WTW analyses of vehicle fuels under localized conditions in China [23–28].

The basic platform in this model was adapted from the GREET model, though it incorporates Chinese characteristics. The platform allows the user to determine the life-cycle fossil fuel intensities and GHG emission intensities of major end-use energies by means of iterative calculations.

In the present study, three primary fossil fuels were considered: coal, petroleum, and natural gas (NG). Nine forms of end-use energy were analyzed, including coal, petroleum, NG-based fuels, and electricity. The life cycle was divided into four phases for analysis: raw material production, raw material transport, fuel production, and fuel transport.

In our analyses, the system boundary was extended to cover the direct use of fuels (i.e., for processing and transport purposes), LCA energy consumption, and GHG emissions. However, indirect energy consumption (e.g., plant infrastructure construction and vehicle manufacture) was ignored.

This model clearly involves circular references. For example, petroleum mining and transportation require diesel, which is itself produced from petroleum. We solved this problem by using the circular reference function (i.e., iterative calculation) offered in Microsoft Excel. Finally, the platform outputs datasheets that detail the LCAs of the energy and carbon-emission intensities of various end-use energies.

In the current study, this model was used to generate the life-cycle primary fossil fuel intensities and major end-use energies in addition to their carbon emission intensities. The resulting datasheets were used as the basis (see Section 3) for subsequent analysis.

BASIC DATA

Energy Data

Petroleum and Refined Products

Table 1 summarizes data related to petroleum mining, transportation, and refining. Table 2 presents details of the transportation and distribution of refined petroleum products. It should be noted that petroleum-refining processes require the use of dry gas, which is not, however, one of the nine end-use energies considered in this study. This situation thus required special consideration: in the present study, dry gas was regarded as a raw material that involves the consumption of no additional primary fossil fuels, and it has a GHG intensity of 65 g/MJ.

Table 1: Energy-efficiency data related to petroleum mining and refining (%)

Item	Value
Proportion of crude oil import	55.4
Extraction efficiency of crude oil	93.0
Energy efficiency of gasoline production	89.1
Energy efficiency of diesel production	89.7
Energy efficiency of fuel oil production	94.0

Data assembled from publications [2, 3] and expert opinions.

Table 2: Data related to petroleum transport and the transport and distribution of refined petroleum products

Transport mode	Crude oil		Gasoline, diesel, and fuel oil	
	Percentage (%)	Average transport distance (km)	Percentage (%)	Average transport distance (km)
Ocean shipping	50	11000	0	0
Train	30	942	50	900
Pipeline	78	440	15	160
Ship	10	250	10	1200
Short-distance vehicle transport	0	0	10	50

Data assembled from publications [2, 3] and expert opinions.
Owing to relay during transport, the sum of data for all modes may exceed 100%.

Natural Gas

Tables 3 and 4 summarize the data related to the extraction, processing, transport, and distribution of natural gas.

Table 3: Energy-efficiency data related to natural gas extraction and processing (%)

Item	Value
Extraction efficiency of natural gas	96.00
Natural gas processing efficiency	94.00

Table 4: Data related to pipeline transport of natural gas.

Application	Average transport distance (km)
Compressed natural gas	625
Process fuel	1500

Carbon Emission-Related Estimations

For our model, the carbon content and oxidation ratios of various types of energy were collected from an authoritative work containing basic data on GHG emissions in China. Methane (CH_4) emission factors for the combustion of oil and gas fuels under different facility conditions were obtained from IPCC report [29] and a report published by the 3E-THU (2003) [30]. The report was supported by the Ministry of Science and Technology of China and completed in collaboration with experts in the petroleum and petrochemical industries. The report estimated the methane emissions associated with various processes involved in the extraction, processing, transport, and consumption of petroleum and natural gas in China. Additionally, methane emission factors for coal combustion under different facility conditions were also collected. Nitrous oxide (N_2O) emission factors under various conditions used in the model were the default values published by IPCC (2006) [29].

It should be noted that natural gas extraction operations involve a small fraction (approximately 0.34%) of methane loss into the atmosphere.

WTW CLASSIFICATION OF PHASES OF NG FUELS

After extraction and purification, NG is delivered to plants, where it is processed into liquid fuels or CNG for vehicle use. Figure 1 summarizes the life-cycle phases of NG fuels.

Figure 1: WTW classification of phases of NG fuels.

This section is primarily concerned with the energy use and GHG emissions of fuels. Specifically, our analyses were focused on two phases: well to pump (WTP) and pump to wheels (PTW). WTP refers to the upstream processes in vehicle fuel production, including resource extraction, resource transport, and the production, transport, distribution, storage, and tank-filling of fuel products. PTW refers to the downstream processes, including primarily the use (combustion) of fuels by vehicles and exhaust emissions.

In our analyses, the system boundary covered the direct use (i.e., process and traffic purposes) of fuels, as well as the LCA energy consumption and GHG emissions. Indirect energy consumption (e.g., plant infrastructure construction and vehicle production) was ignored. However, our analyses considered methane loss (i.e., leakage to the atmosphere) during the production of coal, raw natural gas, and crude oil.

ENERGY CONSUMPTION AND CARBON EMISSIONS DURING VARIOUS PHASES

NG Extraction and Processing

In a previous study, CATARC (2007) [3] analyzed four major NG fields of PetroChina (the fields account for 70% of all natural gas productivity in China) and predicted that the energy efficiency for NG extraction would be approximately 96.4% in 2015. Given the limited variation in this energy efficiency, we set it at 96% in our analyses. Additionally, Shen et al. (2012) [22] reported that the energy consumption for NG extraction and processing (i.e., purification) in China was approximately 10%. Accordingly, the energy efficiency of natural processing (i.e., purification) was estimated to be 94%; that is, (100%–10%)/96% = 94%. Meanwhile, NG extraction and processing primarily consume NG and electricity; the latter is usually provided by gas-fired power stations operated by the same NG fields and processing plants.

NG contains mainly methane, a major GHG. Consequently, methane leakage during NG extraction and processing contributes substantially to GHG emissions. The level of this leakage affects the energy efficiency and emission-reduction performance of NG fuels as alternative vehicle fuels.

CNG Production

CNG fueling stations mainly use electricity, water, and other resources. Additionally, they involve NG loss in the emptying of containers or pipelines. According to CATARC (2007) [3], the energy efficiency for NG compression is 96.9%, and this process involves the loss of NG (0.038 MJ/m^3 NG) and other energy sources (0.56 MJ/m^3 NG).

LNG Production

LNG-related processes primarily include liquefaction plants and LNG reception stations. The energy consumption involved in these processes includes electricity, water, and other resources. For imported LNG, the related energy consumption and efficiency were obtained from the GREET model. The comprehensive energy consumption was determined to be 90.2%, which included predominantly NG (98%) and a small fraction of electricity (2%). For LNG produced in China, there were two technical possibilities: (1) liquefaction near the gas field followed by short-distance transport and distribution for vehicle use; (2) long-distance pipeline transport of NG followed by liquefaction and short-distance transport and distribution for vehicle use. After consultation with the China National Offshore Oil Corp., we set the comprehensive energy efficiency for LNG production in China at 95%. LNG production consumes primarily electricity.

Gas-to-Liquid (GTL) Fuel Production

Currently, there is no GTL fuel production in China, and data related to the energy use of this process were obtained from the GREET model. The comprehensive energy consumption for this process was set at 54.2%. This process consumes predominantly NG (99.7%) and a minor fraction (0.3%) of n-butane.

NG Transport and Product Fuel Transport and Distribution

After purification, NG is generally transported by a long-distance pipeline to reach industrial users or city hubs. After delivery to city hubs, NG is further transported and distributed to final users. In recent years, the average distance of NG pipeline transport in China has rapidly increased—from 217 km in 1998 and 496 km in 2004 to 800 km in 2005. Moreover, several long-distance pipelines have been constructed, such as the west-east natural gas transmission engineering (3,900 km) and the China-Kazakhstan pipeline (3,000 km). In light of this trend, the average distance of NG pipeline transport was set at 1,500 km.

Because WTW analysis of NG-based vehicle fuels is sensitive to the mode and distance of NG transport, different transport distances were set for CNG, LNG, and GTL. The transport distance for CNG production is usually the average distance for NG transport within the country. Since CNG vehicles are predominantly used currently in regions with rich NG resources, the typical pipeline transport distance is approximately 300 km (e.g., the distance between the Kaixian gas field and Chongqing), and subsequent delivery is accomplished by vehicle transport.

Similar to earlier analyses (Section 5.3), three routes of NG supply application were considered: overseas import; local liquefaction followed by vehicle transport; pipeline transport followed by liquefaction and subsequent transport and distribution for vehicle use. For the first route, the average transport distance was set at 6,700 km (Shen et al., 2012) [22]. For the second route, the average vehicle transport distance was assumed to be 100 km. For the third route, the distance for transport and distribution was also set at 100 km.

For GTL fuels, an appropriate means of supply and application consists of production near gas fields followed by transport and distribution to final users. In this study, the typical distance between the gas field and the production plant was assumed to be 100 km, and the subsequent transport and distribution were assumed to be similar to those for diesel.

Moreover, the following assumption was made for estimating the energy consumption for NG transport: pipeline transport is driven by gas turbines that consume mainly NG (90%) and external electricity (10%) from the power grid. Data related to the energy consumption for overseas shipment of LNG were obtained from the GREET model.

Carbon Emissions during Various Phases of NG-Based Fuel Production and Use

In this section, carbon-emission patterns associated with the production and use of various phases of NG-based fuels (i.e., NG, CNG, LNG, and GTL) are analyzed.

Natural Gas

According to our calculations, the upstream processes account for 14.1% of the life-cycle GHG emissions of NG production and use: this comprises 12.6% emitted during NG extraction and processing and 1.5% by subsequent transport. Combining the GHG emissions during upstream processes and final applications, the NG energy chain produces total carbon emissions of 67.1 g $CO_{2,e}$/MJ.

CNG

For CNG, the upstream processes account for 21.2% of the life-cycle GHG emissions. Of these, NG extraction, NG transport, NG compression, and CNG transport amount to 11.6%, 0.3%, 9.3%, and 0.0%, respectively, of the total GHG emissions. In combination with GHG emissions from upstream processes and final application, this energy chain totally emits 73.2 g $CO_{2,e}$/MJ.

LNG Fuel

As described earlier, three supply routes were considered (Section 5.5). Our calculations suggest that for the first route (overseas import followed by supply to local cities for vehicle use), the upstream processes amount to 24.0% of the life-cycle GHG emissions: this comprises 11.2% from natural extraction, 9.4% from NG compression, 2.6% from LNG transport, and 0.8% from LNG distribution. Under this route, the LNG energy chain totally emits 75.7 g $CO_{2,e}$/MJ.

For the second route (liquefaction near gas fields followed by truck transport for vehicle use), the upstream processes account for 25.7% of the life-cycle GHG emissions: this comprises 10.9% from NG extraction, 14.0% from liquefaction, and 0.8% from LNG transport and distribution. Combining upstream processes and final application, the energy chain under this route totally emits 77.5 g $CO_{2,e}$/MJ.

For the third route (NG pipeline transport followed by liquefaction and further transport and distribution for vehicle use), the upstream processes account for 26.7% of the life-cycle GHG emissions: this comprises 10.8% from NG extraction and processing, 1.3% from NG transport and distribution, 13.8% from liquefaction, and 0.8% from LNG transport and distribution. Combining upstream processes and final application, the energy chain under this route totally emits 78.5 g $CO_{2,e}$/MJ.

GTL Fuel

For the application of GTL as an alternative vehicle fuel, the upstream processes amount to 48.0% of the life-cycle GHG emissions: this comprises 7.7% from NG extraction, 0.1% from NG transport, 40.1% from GTL production, and 0.2% from GTL transport and distribution. Combining upstream processes and final use, the GTL energy chain totally emits 146.8 g $CO_{2,e}$ /MJ.

LIFE-CYCLE GHG EMISSION: ANALYSIS AND SUMMARY

Using a medium-size passenger car with an energy efficiency of 8 liters of gasoline consumed per 100 km as the baseline model, we can calculate the WTW fossil energy input and GHG emissions of such pathways as those for gas-based fuels. For different vehicle and fuel technology pathways, the fuel economy situation is presented in ratios using gasoline spark ignition (SI) vehicles as the baseline: diesel is 110%, CNG is 95.0%, LNG is 99.1%, and GTL is 110%. It should be noted that the vehicles are gauged under hypothetical conditions with heating and air-conditioning in use. The fuel consumption in real operating conditions is about 15% higher than in laboratory tests for inner combustion engine (ICE) vehicles.

Figures 2 and 3 summarize the GHG emission behavior of various NG-based fuels for vehicles. CNG- and LNG-powered vehicles have similar WTW fossil energy uses to conventional gasoline- and diesel-fueled vehicles, but differences emerge with the distance of NG transportation. Additionally, thanks to NG having a lower carbon content than petroleum, CNG- and LNG-powered vehicles emit 10–20% and 5–10% less GHGs than gasoline- and diesel-fueled vehicles, respectively. We assumed that errors in our modeling of CNG and LNG fuels would mainly arise with respect to the actual distance of transport.

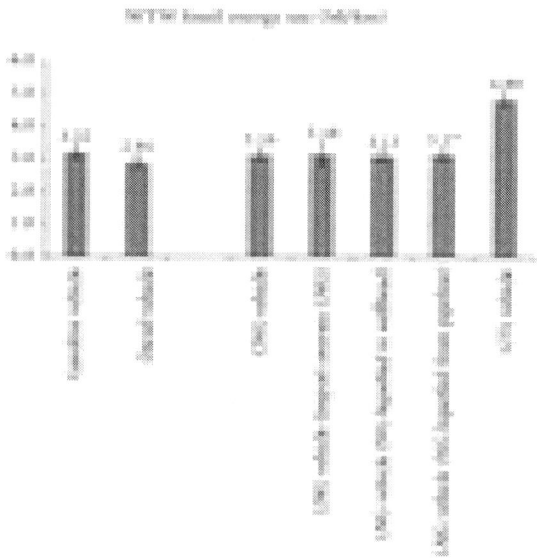

Figure 2: WTW fossil energy use of vehicles powered by NG-based fuels (MJ/km).

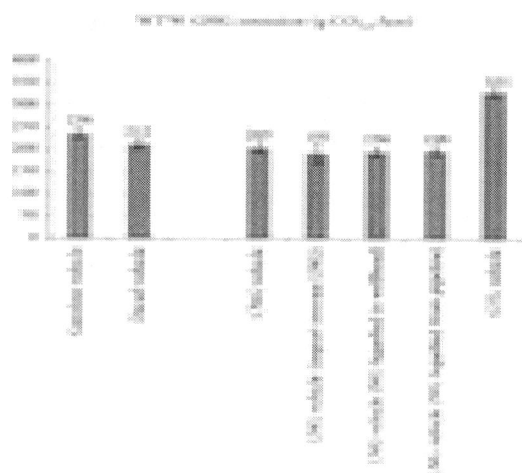

Figure 3: WTW GHG emission levels of vehicles powered by NG-based fuels (g $CO_{2,e}$/km).

However, GTL-powered vehicles involve approximately 50% more WTW fossil energy use than conventional gasoline- and diesel-fueled vehicles, primarily because of the low efficiency of GTL production. Nevertheless, since NG has a lower carbon content than petroleum, GTL-powered vehicles emit approximately 30% more GHGs than conventional-fuel vehicles. We considered uncertainty over the efficiency of future GTL production technology as representing a significant area of error in our analyses.

Since LNG is primarily targeted at commercial vehicles, we made efforts to compare the application of this alternative fuel for heavy trucks and buses. However, because of the lack of consistent data, it was difficult to make an accurate comparison. According to a study by Tang et al. (2011), the replacement ratio for diesel to NG is 90:100 (in terms of net energy). However, according to another source (http://wenku.baidu.com/view/75a1357e168884868762d6ba.html###), the diesel-to-NG replacement ratio should be 78:100 (in net energy). Because of this inconsistency, we performed calculations using these two values as the upper and lower limits. We found that, depending on the replacement ratio, LNG-powered vehicles may emit either 5% less or 12% more GHGs than conventional-fuel vehicles.

SENSITIVITY OF CARBON FOOTPRINT OF LNG

This section analyzes the sensitivity of the carbon footprint of LNG as an alternative vehicle fuel.

Effect of LNG Supply Routes

As mentioned earlier (Section 5.5), three LNG supply routes were considered in this study: (1) overseas import (average shipment distance, 6,700 km (Shen et al., 2012) followed by supply to local cities (average transport distance, 100 km); (2) liquefaction near gas fields followed by truck transport (average distance, 100 km) for final vehicle use; (3) NG pipeline transport followed by liquefaction, transport, and distribution (distance, 100 km) for vehicle use. Our analyses suggest that the first route involves the lowest carbon intensity (75.7 g $CO_{2,e}$/MJ) and the second route the highest carbon intensity (78.5 g $CO_{2,e}$/MJ). The third route results in intermediate carbon intensity of 77.5 g $CO_{2,e}$/MJ.

Impact of Energy Efficiency and Energy Sources in NG Liquefaction

The carbon intensity of the LNG energy chain varied substantially with the type of energy used for NG liquefaction. Our calculations suggest that under the first supply route, assuming that the liquefaction plant uses electricity at a comprehensive energy efficiency of 95.2%, the upstream processes contribute 27.5% of the life-cycle GHG emissions. Combining these upstream processes and the subsequent LNG application, the LNG energy chain totally emits 79.3 g $CO_{2,e}$/MJ. This value amounts to a 4.8% higher carbon intensity than with another assumed situation (liquefaction primarily using NG).

Under the second route, assuming that the plant uses primarily NG at a comprehensive energy efficiency of 90.2%, the upstream processes contribute 21.6% of the life-cycle GHG emissions. Combining these upstream processes and final application, the LNG energy chain is expected to emit a total 73.5 g $CO_{2,e}$/MJ. This value amounts to a 5.2% lower carbon intensity than with another situation (liquefaction using electricity).

Moreover, with the third route, assuming that the liquefaction plant uses primarily NG at a comprehensive energy efficiency of 90.2%, the upstream processes contribute 22.7% of the life-cycle GHG emissions. Considering the upstream processes and final application, this energy chain totally emits 74.4 g $CO_{2,e}$/MJ. This emission level results in a 5.2% lower carbon intensity than with another assumed operation situation (liquefaction using electricity).

Effects of NG Transport

Under the three supply modes, reducing the transport distance for NG and LNG by 50% decreased the carbon intensity as follows: from 75.7 g/MJ to 74.5 g/MJ (first mode, 1.7% decrease); from 77.5 g $CO_{2,e}$/MJ to 77.2 g $CO_{2,e}$/MJ (second mode, 0.4% decrease); from 78.5 g/MJ to 77.7 g $CO_{2,e}$/MJ (third mode, 1.1% decrease).

Summary for the Sensitivity Analysis

Our calculations suggest that the carbon emission intensity of the LNG energy chain is highly sensitive to the efficiency of NG liquefaction and the form of energy used in that process. Moreover, this carbon emission intensity is moderately sensitive to different feedstock supply and fuel-production pathways but relatively insensitive to the distance of NG transport and LNG transport and distribution.

CONCLUDING REMARKS

- CNG- and LNG-powered vehicles have similar WTW fossil energy uses to conventional gasoline- and diesel-fueled vehicles, but differences emerge with the distance of NG transportation. Additionally, thanks to NG having a lower carbon content than petroleum, CNG- and LNG-powered vehicles emit 10–20% and 5–10% less GHGs than gasoline- and diesel-fueled vehicles, respectively.
- However, GTL-powered vehicles involve approximately 50% more WTW fossil energy uses than conventional gasoline- and diesel-fueled vehicles, primarily because of the low efficiency of GTL production. Nevertheless, since NG has a lower carbon content than petroleum, GTL-powered vehicles emit approximately 30% more GHGs than conventional-fuel vehicles.
- The carbon emission intensity of the LNG energy chain is highly sensitive to the efficiency of NG liquefaction and the form of energy used in that process.

ACKNOWLEDGMENTS

The project is cosupported by the China National Natural Science Foundation (Grant no. 71041028, 71103109, and 71073095), China National Social Science Foundation (09 & ZD029), MOE Project of Key Research Institute of Humanities and Social Sciences at universities in China (2009JJD790029), and the CAERC program (Tsinghua/GM/SAIC-China).

REFERENCES

1. CAERC (China Automotive Energy Research Center, Tsinghua University), China Automotive Energy Outlook 2012, Scientific Press, Beijing, China, 2012.
2. A. L. Zhang, W. Shen, W. J. Han, and Q. H. Chai, Life Cycle Analysis of Automotive Alternative Fuel, Tsinghua University Press, Beijing, China, 2008.
3. CATARC (China Automotive Technology and Research Center), Well-to-Wheels Analysis of Energy Consumption and GHG Emissions of Multi Vehicle Fuel in Future China, CATARC, Beijing, China, 2007.
4. NGV Global (natural gas vehicle global), "China's NGV Growth Accelerating," 2012, http://www.iangv.org/category/country/china/.
5. CLNGVN (China LNG Vehicle Net), "LNG buses soar up in China," 2012, http://www.lngche.com/article-1396-1.html.
6. M. A. Delluchi, Emissions of GHG from the Use of Transportation Fuels and Electricity-Volume1: Main Text, Center for Transportation Research, Argonne National Laboratory, Lemont, Ill, USA, 1991.
7. M. A. Delluchi, A Lifecycle Emissions Model (LEM): Main Texts, Institute of Transportation Studies. University of California, Davis, Calif, USA, 2003.
8. M. Wang, GREET 1. 5-Transportaion Fuel-Cycle Model-Volume 1: Methodology, Development, Uses, and Results, Center for Transportation Research, Argonne National Laboratory, Lemont, Ill, USA, 1999.
9. M. Wang, H. Lee, and J. Molburg, "Allocation of energy use in petrol refineries to petrol products," International Journal of Life Cycle Assessment, vol. 9, no. 1, pp. 34–44, 2004.
10. Concawe, EUCAR and EC Joint Research Centre, "Well-to-wheels analysis of future automotive fuels and powertrains in the European context," 2007, http://iet.jrc.ec.europa.eu/about-jec/sites/iet.jrc.ec.europa.eu.about-jec/files/documents/wtw3_wtw_report_eurformat.pdf.
11. J. Wallace, M. Wang, T. Weber, and A. Finizza, "GM study: well-to-wheel energy use and greenhouse gas emissions of advanced

fuel/vehicle systems—North American analysis," Tech. Rep., 2001, http://greet.es.anl.gov/publication-3plz9fyi.

12. M. A. Weiss, J. B. Heywood, E. M. Drake, A. Schafer, and F. F. AuYeung, "On the road in 2020," Tech. Rep. MIT EL 00-003, Laboratory for Energy and the Environment (LFEE), Cambridge, Mass, USA, 2000.

13. M. A. Weiss, J. B. Heywood, A. Schafer, and V. K. Natarajan, "Comparative assessment of fuel cell cars," Tech. Rep. MIT LFEE, 2003-001 RP, Laboratory for Energy and the Environment (LFEE), Cambridge, Mass, USA, 2003.

14. J. Ally and T. Pryor, "Life-cycle assessment of diesel, natural gas and hydrogen fuel cell bus transportation systems," Journal of Power Sources, vol. 170, no. 2, pp. 401–411, 2007.

15. D. Karman, "Life-cycle analysis of GHG emissions for CNG and diesel buses in Beijing," in Proceedings of IEEE EIC Climate Change Technology Conference (EICCCC '06), pp. 1–6, May 2006.

16. L. Kliucininkas, J. Matulevicius, and D. Martuzevicius, "The life cycle assessment of alternative fuel chains for urban buses and trolleybuses," Journal of Environmental Management, vol. 99, pp. 98–103, 2012.

17. F. Ryan and B. Caulfield, "Examining the benefits of using bio-CNG in urban bus operations," Transportation Research D, vol. 15, no. 6, pp. 362–365, 2010.

18. L. Rose, M. Hussain, S. Ahmed, K. Malek, R. Costanza, and E. Kjeang, "A comparative life cycle assessment of diesel and compressed natural gas powered refuse collection vehicles in a Canadian city," Energy Policy, vol. 52, pp. 453–461, 2013.

19. A. Arteconi, C. Brandoni, D. Evangelista, and F. Polonara, "Life-cycle greenhouse gas analysis of LNG as a heavy vehicle fuel in Europe," Applied Energy, vol. 87, no. 6, pp. 2005–2013, 2010.

20. CARB (California Air Resource Board), "Comparison of greenhouse gas emissions from natural gas and diesel vehicles," Tech. Rep. 08/10/2008.CARB, 2008.

21. O. P. R. van Vliet, A. P. C. Faaij, and W. C. Turkenburg, "Fischer-Tropsch diesel production in a well-to-wheel perspective: a carbon, energy flow and cost analysis," Energy Conversion and Management, vol. 50, no. 4, pp. 855–876, 2009.

22. W. Shen, W. J. Han, D. Chock, Q. H. Chai, and A. L. Zhang, "Well-to-wheels life-cycle analysis of alternative fuels and vehicle technologies in China," Energy Policy, vol. 49, pp. 296–307, 2012.
23. X. Li, X. Ou, X. Zhang, Q. Zhang, and X. L. Zhang, "Life-cycle fossil energy consumption and greenhouse gas emission intensity of dominant secondary energy pathways of China in 2010," Energy, vol. 50, pp. 15–23, 2013.
24. X. Ou, Y. Xiaoyu, and X. Zhang, "Life-cycle energy consumption and greenhouse gas emissions for electricity generation and supply in China," Applied Energy, vol. 88, no. 1, pp. 289–297, 2011.
25. X. Ou, X. Zhang, S. Chang, and Q. Guo, "Energy consumption and GHG emissions of six biofuel pathways by LCA in China," Applied Energy, vol. 86, no. 1, pp. S197–S208, 2009.
26. X. Ou, X. Zhang, and S. Chang, "Alternative fuel buses currently in use in China: life-cycle fossil energy use, GHG emissions and policy recommendations," Energy Policy, vol. 38, no. 1, pp. 406–418, 2010.
27. X. Ou, X. Zhang, and S. Chang, "Scenario analysis on alternative fuel/vehicle for China›s future road transport: life-cycle energy demand and GHG emissions," Energy Policy, vol. 38, no. 8, pp. 3943–3956, 2010.
28. X. M. Ou and X. L. Zhang, Life-Cycle Analysis of the Automotive Energy Pathways in China, Tsinghua University Press, Beijing, China, 2011.
29. IPCC (Intergovernmental Panel on Climate Change), "IPCC Guidelines for National GHG Inventories," 2006.
30. 3E-THU (Institute of Energy, Environment and Economy at Tsinghua University), Report of Inventory of CH4 Emission Sources and Sinks of China's Petroleum and Natural Gas Sector in 2000, Institute of Energy, Environment and Economy, Tsinghua University, Beijing, China, 2003.

Chapter 5

Study on Nonequilibrium Effect of Condensate Gas Reservoir with Gaseous Water under HT and HP Condition

Dali Hou, Pingya Luo, Lei Sun, Yong Tang, and Yi Pan

The State Key Laboratory of Oil & Gas Reservoir Geology and Exploitation Engineering, Southwest Petroleum University, Xindu Avenue No. 8, Chengdu, Sichuan Province 610500, China

ABSTRACT

When a condensate gas reservoir with gaseous water under high temperature and high pressure condition is producing, the gaseous water and nonequilibrium effect will have great influences on the phase behavior of condensate oil and gas system and the accumulation of condensate liquid near the wellbore area. Therefore, a series of experiments were performed to investigate phase behavior of the condensate gas reservoirs with gaseous water using a PVT cell, in which the constant volume depletion process of nonequilibrium pressure drop and equilibrium pressure drop within near wellbore zone was simulated. And using the modified PR EOS, PR EOS, and nonequilibrium effect theory, the authors calculated the content of condensate oil and condensate liquid of the nonequilibrium pressure drop and equilibrium pressure drop and compared the calculated results with the experimental data. The results show that the modified PR EOS combined with nonequilibrium effect theory is more suitable for representing phase behavior characteristics of the development process of condensate gas reservoir containing gaseous water, with the average relative error of 4.49%. Furthermore, choosing the appropriate exploiting opportunity and properly increasing the nonequilibrium effect are helpful to increase condensate oil and water recovery.

INTRODUCTION

Condensate reservoir with gaseous water is an important part of oil-gas resources [1–3]. Studies have shown that the influence of formation water on condensate gas phase behavior cannot be ignored. The reason is that water in the gas phase mixes with gaseous hydrocarbon under high temperature and high pressure condition, and the polarity of the water molecule leads to a change in the phase behavior of condensate gas in the systems that contain gaseous water [4–9]. Because the water molecule is polar, the conventional hydrocarbon system equation of state is no longer adapted to describe the condensate gas reservoirs with gaseous water. Nichita et al. predicted the phase equilibrium data by using the modified Peng-Robinson EOS, which utilized the different binary interaction parameters in the aqueous and hydrocarbon gases [5]. Oliveira et al. utilized the CPA EOS to accurately describe the mutual solubilities of water and several aliphatic and aromatic hydrocarbons in a broad range of pressures and temperatures [6]. Pedersen and Milter measured phase equilibrium data for the mutual solubility of brine and a gas condensate mixture at temperatures ranging from 35°C to 200°C and pressures of 70 MPa to 100 MPa [7]. These experimental data were modeled using the PR EOS with the nonclassical mixing rule of Huron and Vidal. The simulation results are in good agreement with the experimental data even at high pressure and high temperature. Lindeloff and Michelsen calculated the pressure-temperature phase diagrams for hydrocarbon-water mixtures using the SRK EOS with temperature dependent Peneloux volume shift [8].

During the development process of condensate gas reservoirs with gaseous water, both reservoir pressure and wellbore pressure decrease gradually. Retrograde condensation takes place as soon as reservoir pressure falls below the dew point of gas condensate with gaseous water in place, which includes the change of single-phase flow into oil-gas-water three-phase flow [10]. There is no time to reach fully phase equilibrium between gas and liquid phase because of a high-speed flow of the gas phase. A part of the precipitation liquid phase moves slowly and gradually settles due to the porous media adsorption effect of the rock surface, which is caused by the remaining free energy of the rock surface. The other part of the precipitation liquid phase entrained in the gas phase moves a distance with high speed and then forms the liquid. This is the so-called "precipitate lag" phenomenon, namely, the nonequilibrium effect [11]. Therefore, the phase behavior analysis method of condensate gas based on the equilibrium theory cannot simulate the real nonequilibrium process of oil and gas system near the wellbore zone [12–16]. Moreover, a conventional phase behavior experiment based on the equilibrium theory needs to consume large amounts of time and the experimental analysis process is not consistent with the actual gas field exploitation [17–19]. Currently, there are few relevant experimental reports on the phase behavior of gas condensate with gaseous water in the process of nonequilibrium pressure drop. Therefore, it is necessary to carry out experimental research on this topic and the experimental results have a very important engineering significance for the actual production of a gas condensate with gaseous water.

In this work, simulation modeling and experimental analysis of nonequilibrium effect in the gas condensate with gaseous water are considered. The main purpose is to compare the phase behavior data considering gaseous water in the process of the nonequilibrium pressure drop with the phase behavior data considering gaseous water in the process of the equilibrium pressure drop.

THEORY

Thermodynamic Model for a Gas-Condensate/Water System

The Peng-Robinson equation of state [20] is one of the most widely used for engineering EOS:

$$P = \frac{RT}{V-b} - \frac{a(T)}{V(V+b) + b(V-b)}. \tag{1}$$

For a pure fluid, constant b is given by

$$b_i = 0.007780 \left(\frac{RT_{ci}}{P_c} \right). \tag{2}$$

While a (T), a function of temperature, is given by

$$a_i(T) = a_i(T_{ci}) \times \alpha_i(T_{ri}, \omega_i),$$

$$a_i(T_{ci}) = 0.45724 \frac{R^2 T_{ci}^2}{P_{ci}},$$

$$\alpha_i(T_{ri}, \omega_i) = \sqrt{1 + m\left(1 - T_{ri}^{1/2}\right)},$$

$$m = 0.37464 + 1.54226\omega_i - 0.26992\omega_i^2. \tag{3}$$

A simple classical Van der Waals mixing rule with one binary interaction parameter k_{ij} is used to calculate mixture systems:

$$a_m(T) = \sum_{i=1}^{n}\sum_{j=1}^{n} x_i x_j \left(a_i a_j \alpha_i \alpha_j\right)^{0.5} (1 - k_{ij}),$$

$$b_m = \sum_{i=1}^{n} x_i b_i. \tag{4}$$

For polar-non-polar interactions, however, the classical Van der Waals mixing rule for the attractive parameter α is not satisfactory and an unconventional form of the classical mixing is required. In this study, Huron and Vidal's modified attractive parameter α is introduced, which combines Van der Waals mixing rule with activity coefficient [7]. The Huron and Vidal's modified attractive parameter α is expressed:

$$a = b\left(\sum_{i=1}^{n}\left(x_i \frac{a_i}{b_i}\right) - \frac{G_\infty^E}{\lambda}\right). \tag{5}$$

λ is a constant, which has different values for different EOS. For PR EOS, it is equal to the following value:

$$\lambda = \frac{1}{2\sqrt{2}} \ln\left(\frac{\sqrt{2}+1}{\sqrt{2}-1}\right). \tag{6}$$

G_∞^E represents Gibbs free energy at infinite pressure, which is computed using modified NRTL mixing rule:

$$\frac{G_\infty^E}{RT} = \sum_{i=1}^{n} x_i \frac{\sum_{j=1}^{n} \tau_{ji} b_j x_j \exp(-\alpha_{ji}\tau_{ji})}{\sum_{k=1}^{n} b_k x_k \exp(-\alpha_{ki}\tau_{ki})}. \tag{7}$$

α_{ij} is not a random parameter; τ_{ij} represents the intermolecular forces:

$$\tau_{ij} = \frac{g_{ji} - g_{ii}}{RT}, \tag{8}$$

Where g_{ij} is the energy characteristic parameters between the component j and component i, which is expressed as a function of temperature. The expression is as follows [7]:

$$g_{ji} - g_{ii} = (g_{ji} - g_{ii})' + T(g_{ji} - g_{ii})''. \tag{9}$$

Here, $(g_{ij}-g_{ii})'$ and $(g_{ij}-g_{ii})''$ have nothing to do with temperature. When $a_{ij}=0$, there is

$$g_{ii} = -\frac{a_i}{b_i}\lambda,$$

$$g_{ji} = -2\frac{\sqrt{b_i b_j}}{b_i + b_j}(g_{ii} g_{jj})^{0.5}(1 - k_{ij}). \tag{10}$$

The Huron-Vidal mixing rules are simplified as classical Van der Waals mixed rules. This model can be used to calculate phase equilibrium parameters for the systems containing polar material, such as the system containing the condensate gas, condensate oil, and gaseous water.

Nonequilibrium Pressure Drop Theory

А.Х. МИРЗАДЖАНЗАДЕ has done a lot of research on the nonequilibrium effect for condensate gas system and proposed a correlation about the relationship between the condensate oil content and pressure change speed in the process of nonequilibrium pressure drop [19]. The correlation is as follows:

$$q = A\left(p_b - p + \int_0^t K(T-\tau)\frac{dp(\tau)}{d\tau}d\tau\right), \tag{11}$$

Where

$$A = \frac{q_{max}}{(p_b - p_{max})},$$

$$K(t) = K_0 e^{-t/T}. \qquad (12)$$

Here, p_b, p_{max} and p, respectively, represent the dew point pressure, the maximum retrograde condensation pressure, and the system pressure; q and q_{max} respectively, represent the yield of retrograde condensate oil at present pressure and the yield of retrograde condensate oil at maximum retrograde condensation pressure, cm³/min; K_0 is the weight coefficient, here, K_0=0.21; t represents the relaxation time, here, T=8800 s; t represents the time in s.

When the pressure change rate is infinitesimal; that is t=T,

$$q = A(p_b - p). \qquad (13)$$

According to the equation of condensate oil content in the nonequilibrium pressure drop, the expression of mole fraction for condensate oil in the instantaneous nonequilibrium phase change can be deduced:

$$L' = L\left(1 + \frac{\int_0^t K(T-\tau)(dp(\tau)/d\tau)d\tau}{(p_b - p)}\right), \qquad (14)$$

where L' represents the mole fraction of condensate oil in the instantaneous nonequilibrium phase change; represents the mole fraction of condensate oil in the equilibrium phase change, which is calculated by the modified PR EOS.

EXPERIMENT

Sample Preparation

The experimental oil and gas samples were directly taken from surface separator in YKL6 well in the gas condensate field of Northwest Oilfield Company, Xijiang, China. The condensate oil content of the YKL6 well is 206 g/m^3. The original formation pressure is 58.72 MPa, the formation temperature is 136.5°C, the stable gas-oil ratio is 4165.7 m^3/m^3, the density of stock tank oil is 0.7405 g/cm^3, and saturation pressure is 56.85 MPa. Furthermore, it is a condensate gas reservoir with gaseous water.

Strictly speaking, this is a synthetic condensate gas sample using the separator gas sample and the separator oil sample in the lab. For more detailed descriptions about synthesis process see SY/T5542-2009 [21]. The measured gas-oil ratio and oil density of synthetic condensate gas sample by two-phase flash experiment are 4207.5 m^3/m^3 and 0.7482 g/cm^3, respectively. It is very close to the abovementioned stable gas-oil ratio and stock tank oil density. It is a suggestion that the synthetic condensate gas sample is reasonably representative; thus the test results can be used to guide the oil field production practice. The flash gas sample and oil sample were analyzed using a HP-6890 gas Chromatograph and an Agilent-7890A oil chromatograph. The precision of components analyses is around ±0.01 (mol%). The measured gas-oil ratio data and the components of flashed gas and oil were then used to calculate the components of reservoir fluid (Table 1). The ion concentrations of formation water were analyzed by high-performance ion-exchange chromatography (HPIC) with two ion-exchange columns (NJ-3A-4A, 250 mm, 4.6 mm; Grace609121268, 100 mm, 4.6 mm) and a suppressed conductivity detection. The minimum detection limit of the HPIC was 0.05 mgkg^{-1}. The ion concentrations analysis result of formation water is shown in Table 2. According to the original water saturation, a certain proportion of formation water was added to the condensate gas sample in the PVT cell to obtain synthetic condensate gas sample with gaseous water. The added formation water and the condensate gas sample were stirred well for 2 hours and reached saturated state. Thereafter a two-phase flash experiment was carried out for the synthetic condensate gas sample with gaseous water under the constant original formation pressure condition. The measured gas-oil ratio of synthetic condensate gas sample with gaseous water by two-phase flash experiment is 4008.5 m³/m³. The flash gas sample and oil sample of the synthetic condensate gas sample with gaseous water were also analyzed using a HP-6890 gas Chromatograph and an Agilent-7890A oil chromatograph. The condensate water mass was measured by electronic balance and then the water vapor content was calculated [22]. According to the water vapor content, the mole fraction of gaseous water was calculated as follows [23].

Table 1: Composition of the reservoir fluid studied

Components	1 mol%	2 mol%
H2O	1.52	0.00
CO2	2.40	2.43
N2	6.05	6.14
C1	80.54	81.76
C2	4.36	4.43
C3	1.59	1.61
C4	0.25	0.25
C4	0.51	0.52
C5	0.17	0.17
C5	0.21	0.21
C6	0.03	0.03
C7	0.08	0.08
C8	0.16	0.16
C9	0.27	0.27
C10	0.32	0.33
C11+	1.57	1.49
Total	100	100

[1]The components of the synthetic condensate gas sample with gaseous water; properties of the C_{11+}: molecular weight, 217.9 (g/mol); density at 293.15, 0.8560 g/cm³.

[2]The components of the synthetic condensate gas sample without gaseous water; properties of the C_{11+}: molecular weight, 211.7 (g/mol); density at 293.15, 0.8499 g/cm³.

Table 2: Properties of formation water

Gas well no.	Formation water		Amount of ions of formation water (mg/L)					
	r1	TDS2						
	g/cm-3	mg/L	HCO-3	Cl-	SO42+	Ca2+	Mg2+	Na+ + K+
YKL6	1.121	176336.3	276.45	107919.37	100	8400.25	967.02	58673.21

¹p is the density of formation water, measured by a densimeter at 293.15 K and atmospheric pressure, and its error is lower than 0.01%.
²TDS is the total dissolved solids of formation water

In the hypothesis the gas is 1 m³, the moles of the gas at 20°C, 0.1 MPa are

$$n_g = \frac{1}{V_m}. \tag{15}$$

Here, the V_m is equal to 24.5.
The gaseous water mole fraction is

$$y_w = \frac{W}{M_w n_g}. \tag{16}$$

The modified component of the synthetic condensate gas sample with gaseous water is expressed as follows:

$$y_{i(\text{modified})} = (1 - y_w) y_i. \tag{17}$$

y_i is the component of reservoir fluid.

Thus, the different components of the synthetic condensate gas sample with gaseous water were obtained. The results are shown in Table 1.

As is shown in Table 1, the mole fraction of gaseous water component in the condensate gas with gaseous water is 1.52%. Therefore the mole fraction of volatile components and gas-oil ratio in the condensate gas with gaseous water are slightly lower than the mole fraction of volatile components and gas-oil ratio in the condensate gas without gaseous water. In contrast, the mole fraction and molecular mass of C_{11+} components are higher in the condensate gas with gaseous water, which suggests that the presence of gaseous water increases the content of C_{11+} components in the condensate gas system and increases the weight of the system.

Apparatus

The experimental facility is a mercury-free high pressure PVT system made by the DBR Company, Canada. A schematic diagram of the corresponding apparatus is given in Figure 1. It is mainly characterized by the visual observation of experimental phenomena and the specially designed piston in the PVT cell that allows to precisely measure even minute liquid. The system is mainly made up of PVT cell, thermostatic air bath, pressure sensor, temperature sensor, sample container, automatic pump, and operation control system. Sample container is made from sapphire glass. The volume of fluid in the PVT cell can be calculated by the internal cross-sectional area of sapphire glass cylinder multiplied by the height of the fluid, which was measured by the grating altimeter.

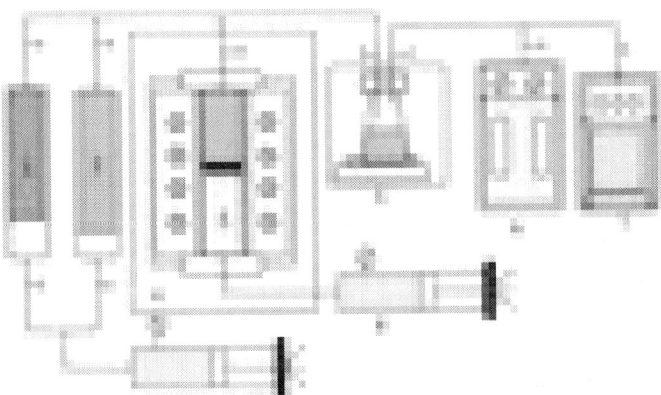

Figure 1: Schematic diagram of the experimental apparatus. 1—water sample, 2—condensate gas sample, 3—PVT cell, 4—thermostatic system, 5—oil

and gas water separation devices and electronic balance, 6—gasometer, 7—oil and gas chromatography, 8—automatic pump.

The physical parameters of main parts are as follows.

- PVT Cells: maximal working pressure is 70 MPa, the highest operating temperature is 200°C, and maximal volume is about 150 cm^3.
- Pressure Sensor: 0~100 MPa, precision of pressure control is 1 Psi
- Temperature Sensor: 0~200°C, precision of temperature control is ±0.1°C.
- Thermostatic air bath: highest operation temperature is 200°C and precision of temperature control is ±0.1°C.
- Sample container: 130 cm^3.
- Automatic pump: rated operation pressure is 100 MPa with resolution of 1 Psi, the range of operation pump volume is 500 cm^3 with resolution of 0.001 cm^3.

The accuracy of PVT main parts satisfies the phase behavior experimental requirement of the synthetic condensate gas sample containing gaseous water.

Experimental Procedure

- Clean PVT cells and then connect the vacuum pump to the PVT cells and evacuate the cells.
- Control and maintain the formation temperature using the thermostatic air bath.
- Introduce a certain proportion of the synthetic condensate gas sample and excess formation water into PVT cells to a formation pressure, adjust the oven through thermostat to the formation temperature; stir the condensate gas sample and formation water for 1 hour and discharge superfluous water sample, and then maintain for 30 min and measure the condensate gas sample volume of the PVT cells.
- Depressurize to the dew point pressure, balance for 1 hour, and record the volume of the sample in the PVT cells. The volume is expressed as V_c (it is used to calculate the condensate oil saturation and the condensate oil and gas recovery).

- It is evenly divided into 6~8 pressure drop between the dew point pressure and the proposed abandonment pressure. Depressurize to the each desirable pressure and balance the system for half an hour or more and then record the total volume of the sample and the volume of the condensate oil in the PVT cell.
- Slightly open the valve between the cells and discharge gas from the PVT cells into the flash separators. When the volume of the sample in the PVT cells is equal to the V_c, the process of discharge gas is over. At the same time, the volume of gas and the mass of the oil and water are, respectively, measured and the components of oil and gas are analyzed using the oil and gas chromatography.
- Repeat steps (4)~(5), until the last level of pressure.
- Slightly open the valve between the cells and depressurize to the atmospheric pressure, discharge gas from the PVT cells into the flash separators, and record the volume of the condensate oil and gas and the volume of condensate water; the flash gas sample and oil sample are analyzed using a HP-6890 gas Chromatograph and an Agilent-7890A oil chromatograph.

The above experiment steps were the equilibrium pressure drop constant volume depletion experiment. In the nonequilibrium pressure drop constant volume depletion experiment, continuously depressurize in a certain rate (dp/dt = 0.007 MPa, dp/dt = 0.028 MPa) in step (5) and then discharge gas from the PVT cells selecting the corresponding pressure in the equilibrium pressure drop constant volume depletion experiment (bleed gas before the pressure get balanced). Record the total volume of the sample in the PVT cells, the volume of condensate oil in the PVT cells, the volume of the condensate oil and gas, and the volume of condensate water in the flash separators. The flash gas sample and oil sample are analyzed using a HP-6890 gas Chromatograph and an Agilent-7890A oil chromatograph.

RESULTS AND DISCUSSION

Analysis for the Condensate Oil Recovery

The condensate oil accumulation recoveries in the equilibrium pressure drop and nonequilibrium pressure drop constant volume depletion experiment are shown in Figure 2. As is shown in Figure 2, the condensate oil recovery under each pressure point in the nonequilibrium pressure drop constant volume depletion experiment is higher than in the equilibrium experiment. The ultimate recovery is 42.55% in the nonequilibrium experiment, which is 6.4% more than the ultimate recovery in the equilibrium experiment. The reason is that the system stability time does not reach the oil and gas separation equilibrium time in the nonequilibrium experiment, which makes the condensate liquid in the form of mist existing in the gas phase and then produces to the ground with natural gas. Therefore, an appropriate nonequilibrium effect in the production process can improve condensate oil recovery.

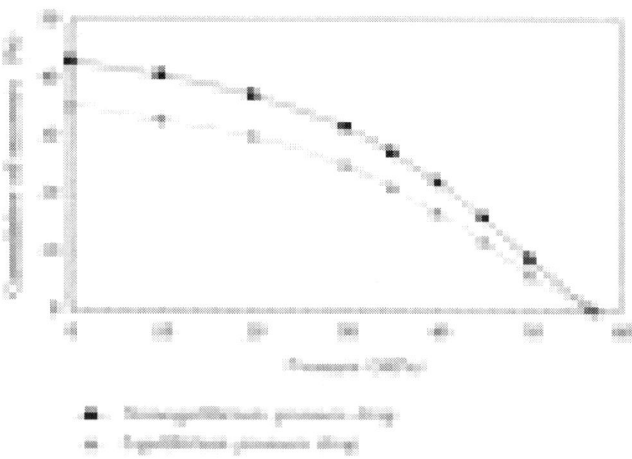

Figure 2: The condensate oil accumulation recovery in the equilibrium pressure drop and nonequilibrium pressure drop constant volume depletion experiment.

The condensate oil recovery under each pressure point in the equilibrium and nonequilibrium experiment is shown in Figure 3. As shown in Figure 3, the condensate oil recovery under each pressure in the nonequilibrium experiment is higher than the condensate oil recovery under each pressure in the equilibrium experiment. Moreover, the condensate oil recovery increases rapidly when the pressure is higher than the maximum retrograde condensation pressure (35 MPa). But the growth rate of the condensate oil recovery in the nonequilibrium experiment is close to the growth rate of the condensate oil recovery in the equilibrium experiment when the pressure is lower than the maximum retrograde condensation pressure (35 MPa). The main reason for this difference is that the condensate oil content is very high when the pressure is higher than the maximum retrograde condensation pressure. During the pressure reduction period the system mainly undergoes a retrograde condensation process which generates more oil condensate during the gas production process in the nonequilibrium experiment. The system mainly undergoes reverse evaporation and the condensate oil content is very low in the condensate gas when the pressure is lower than the maximum retrograde condensation pressure. At this time the produced condensate oil during the gas production process mainly extracts and evaporates the retrograde condensation oil. Therefore, the condensate oil recovery under each pressure point in the nonequilibrium experiment is close to the condensate oil recovery under each pressure in the equilibrium experiment in the later period of constant volume depletion experiment. It is also a suggestion that in the middle and later periods of condensate gas field development, the nonequilibrium effect is weak.

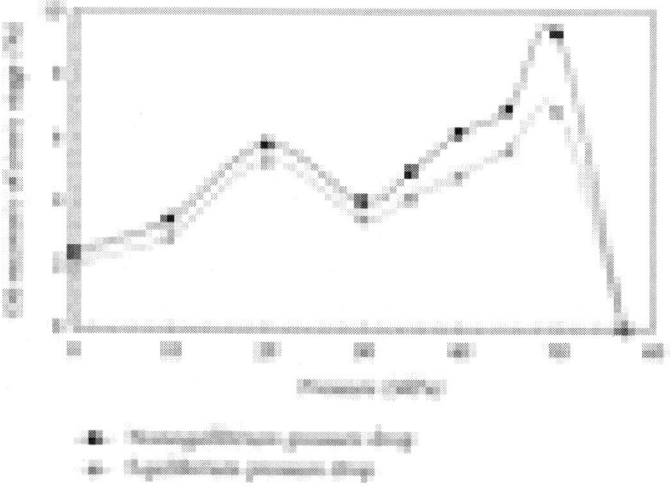

Figure 3: The condensate oil recovery under each pressure point in the equilibrium pressure drop and nonequilibrium pressure drop constant volume depletion experiment.

Analysis for the Condensate Water

In the constant volume depletion experiment, when pressure reduces to 10 MPa, the content of produced condensate water in equilibrium and nonequilibrium pressure drop is, respectively, 20.72 kg/10^3m^3 and 36.67 kg/10^3m^3. And when the pressure reduces to the atmospheric pressure, the content of produced condensate water in equilibrium and nonequilibrium pressure drop is, respectively, 250.95 kg/10^3m^3 and 311.90 kg/10^3m^3. Those data show that the content of produced condensate water in nonequilibrium pressure drop is higher than that in equilibrium pressure drop under the different pressure. The amount of produced condensate water is mainly related to the stabilization time. In the nonequilibrium experiment, the stabilization time does not reach balanced time, so liquid water exists in gas phase in the form of small droplets and is produced together with the gas. That is why the content of produced condensate water in nonequilibrium pressure drop is higher than that in equilibrium pressure drop. Moreover, the experimental result shows that the content of produced condensate water in the low pressure stage is far higher than that of high pressure stage. This happens because the differential pressure near the wellbore zone is large during the production process of condensate gas reservoirs and the well bottom hole pressure is low, which causes the gas effective permeability to drop sharply and the gas well production capacity to correspondingly decrease. This is the famous "reverse dialysis water lock phenomenon." So it is vitally important to prevent the condensate water from undergoing the reversing osmosis water lock near the wellbore area in the low pressure stage.

Modified PR EOS Adaptability Analysis

The average relative error is defined:

$$\varepsilon = \frac{1}{n}\sum_{i=1}^{n}\left|\frac{S_{ocal}}{S_{o\exp}} - 1\right| \times 100\%, \tag{18}$$

Where ε is the average relative error; S_{ocal} is the calculated retrograde condensate oil saturation, decimal; S_{oexp} is the measured retrograde condensate oil saturation, decimal; n is the number of experimental points.

The condensate oil saturation calculated by conventional PR equation, modified PR equation considering the gaseous water and the measured condensate oil saturation in the equilibrium experiment are shown in Figure 4. As is shown in Figure 4, the condensate oil saturations calculated by modified PR EOS considering the gaseous water are close to the measured condensate oil saturation. According to (18), the average relative errors of conventional PR equation and modified PR EOS are, respectively, 12.83% and 4.49%. It is a suggestion that the modified PR equation is more suitable for representing the phase behavior characteristics of condensate gas system with gaseous water under the high temperature and high pressure condition.

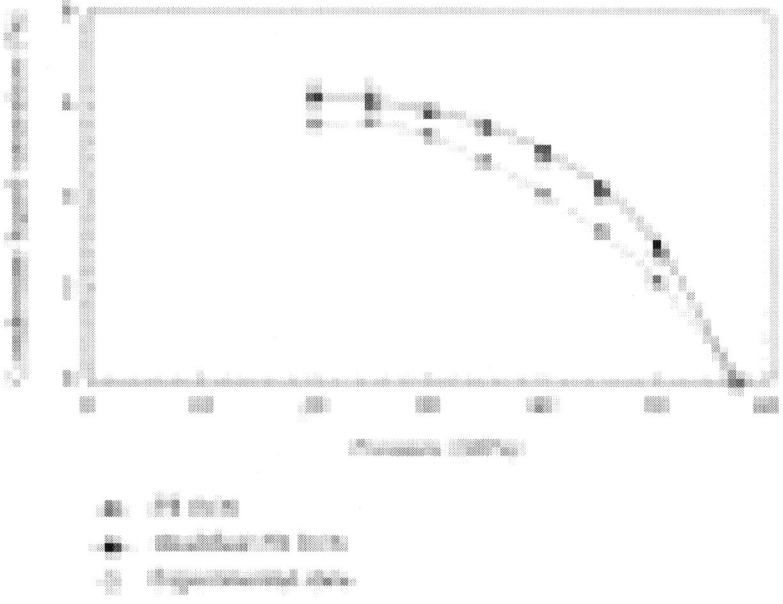

Figure 4: The contrast curve about the calculated retrograde condensate oil saturation in the equilibrium pressure drop by the modified PR EOS and PR EOS, the measured retrograde condensate oil saturation in the equilibrium pressure drop constant volume depletion experiment.

According to the calculated retrograde condensate oil saturation in the process of equilibrium pressure drop by modified PR EOS, we calculated the retrograde condensate oil saturation at different pressure drop rates ($dp/dt = 0.007$ MPa, $dp/dt = 0.028$ MPa) in the nonequilibrium pressure drop using (15). The curve of the pressure versus the calculated retrograde condensate oil saturation is shown in Figure 5.

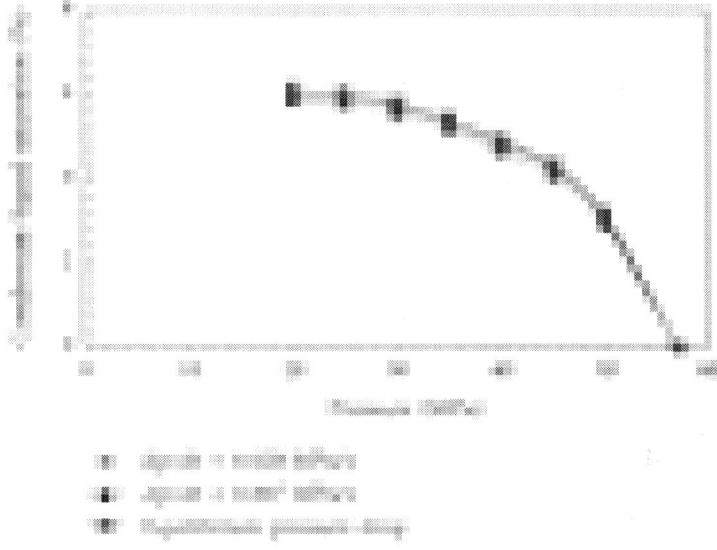

Figure 5: The contrast curve about the calculated retrograde condensate oil saturation in the equilibrium pressure drop by the modified PR EOS, the calculated retrograde condensate oil saturation of different pressure drop rate (dp/dt = 0.007MPa, dp/dt = 0.028 MPa) in the nonequilibrium pressure drop using (15).

As is shown in Figure 5, the more serious is the nonequilibrium effect and the greater is the pressure drop rate, the greater the retrograde condensate oil saturation of the nonequilibrium pressure drop deviates from the retrograde condensate oil saturation of the equilibrium pressure drop. Moreover, the greater the degree of pressure drop, the more the cumulative condensate oil can be produced in the nonequilibrium pressure drop and the lower the retrograde condensate oil saturation is in the formation. Therefore, it can be concluded that the nonequilibrium effect can reduce the retrograde condensate oil saturation in the near wellbore zone and have a certain positive role in slowing retrograde condensation pollution in the near wellbore zone.

CONCLUSIONS

- The condensate oil precipitation lags behind because of the nonequilibrium effect. This part that delayed will be produced to the ground with natural gas. So in the process of production of condensate gas reservoir with gaseous water, the appropriate nonequilibrium effect can increase condensate oil and condensate water recovery.
- When the system pressure is higher than the maximum retrograde condensation pressure, condensate oil content is higher and the nonequilibrium effect is stronger. So in the different stages of condensate gas field exploitation, the degree of nonequilibrium effect is different. Therefore by choosing an appropriate exploiting opportunity and properly increasing the nonequilibrium effect an efficient production of a condensate gas reservoir can be achieved.
- The average relative error of the retrograde condensate oil saturation calculated by modified PR equation is 4.49%. It is a suggestion that the modified PR equation is more suitable for representing the phase behavior characteristics of condensate oil and gas system with gaseous water under the high temperature and high pressure condition.
- In the different stages of condensate gas field with gaseous water exploitation, the nonequilibrium effect has different influences on the condensate oil and condensate water. In other words, nonequilibrium effect has a greater influence on condensate oil in the high pressure stage but has a greater influence on condensate water in the low pressure stage. Therefore, choosing an appropriate exploitation opportunity and properly increasing the nonequilibrium effect are helpful to slow retrograde condensation pollution in the near wellbore area and reverse osmosis water locking damage degree and improve the production efficiency of condensate gas reservoir with gaseous water.

ACKNOWLEDGMENTS

The authors wish to acknowledge anonymous reviewers for constructive comments and suggestions for improving this paper. The authors also wish to thank the anonymous Associate Editor for his handling of the paper and additional suggestions. This work was supported by National Science and Technology Major Project of China (no. 2011ZX05016-005-2) and a Grant from the National Natural Science Foundation of China (no. 50604011).

REFERENCES

1. Y. Li, B. Li, Y. Hu et al., "Water production analysis and reservoir simulation of the Jilake condensate gas field," Petroleum Exploration and Development, vol. 37, no. 1, pp. 89–93, 2010.
2. S. Zendehboudi, M. A. Ahmadi, L. James, and I. Chatzis, "Prediction of condensate-to-gas ratio for retrograde gas condensate reservoirs using artificial neural network with particle swarm optimization,"Energy & Fuels, vol. 26, pp. 3432–3447, 2012.
3. S. A. Jokhio, D. Tiab, and F. Escobar, "Forecasting liquid condensate and water production in two-phase and three-phase gas condensate systems," in Proceedings of the SPE Annual Technical Conference and Exhibition, San Antonio, Tex, USA, September-October 2002, SPE 77549.
4. L. F. Ayala and J. P. Kouassi, "The similarity theory applied to the analysis of multiphase flow in gas-condensate reservoirs," Energy & Fuels, vol. 21, no. 3, pp. 1592–1600, 2007. · ·
5. D. V. Nichita, D. Broseta, P. Elhorga, and F. Montel, "Pseudo-component delumping for multiphase equilibrium in hydrocarbon-water mixtures," Petroleum Science and Technology, vol. 26, no. 17, pp. 2058–2077, 2008. · ·
6. M. B. Oliveira, J. A. P. Coutinho, and A. J. Queimada, "Mutual solubilities of hydrocarbons and water with the CPA EoS," Fluid Phase Equilibria, vol. 258, no. 1, pp. 58–66, 2007. · ·
7. K. S. Pedersen and J. Milter, "Phase equilibrium between gas condensate and brine at HT/HP conditions," in Proceedings of the SPE Annual Technical Conference and Exhibition, Houston, Tex, USA, September 2004, SPE 90309.
8. N. Lindeloff and M. L. Michelsen, "Phase envelope calculations for hydrocarbon-water mixtures," inProceedings of the SPE

Annual Technical Conference and Exhibition, San Antonio, Tex, USA, September-October 2002, SPE 77770.

9. S. KoKal, A. D. Mohammad, and S. Sayegh, "Phase behavior of a gas condensate/water system," in Proceedings of the SPE Annual Technical Conference and Exhibition, Dallas, Tex, USA, October 2000, SPE 62931.

10. K. L. Wu, X. F. Li, and H. T. Wang, "A quantitative model for evaluating the impact of volatile oil non-equilibrium phase transition on degassing," Petroleum Exploration and Development, vol. 39, pp. 636–643, 2012.

11. F. Civan, "Including nonequilibrium relaxation in models for rapid multiphase flow in wells," SPE Production and Operations, vol. 21, no. 1, pp. 98–106, 2006.

12. G. E. Paredes, A. V. Rodríguez, and E. G. Martíneza, "A numerical analysis of non-equilibrium thermodynamic effects in an oil field: two-equation model," Petroleum Science and Technology, vol. 31, pp. 192–203, 2013.

13. G. G. Michel and F. Civan, "Modeling nonisothermal rapid multiphase flow in wells under nonequilibrium conditions," SPE Production and Operations, vol. 23, no. 2, pp. 125–134, 2008.

14. X. D. Kang, X. F. Li, and Y. J. Liu, "Influence on well productivity of high-velocity flow in gas condensate reservoirs," in Proceedings of the 2nd International Symposium on Multiphase, Non-Newtonian and Reacting (Flows '04), vol. 2, pp. 362–366, Hangzhou, China, 2004.

15. F. Civan, "Modeling well performance under nonequilibrium deposition conditions," in Proceedings of the SPE Production and Operations Symposium, Oklahoma City, Okla, USA, March 2001, SPE 67234.

16. W. J. Wu, P. Wang, and M. Delshad, "Modeling non-equilibrium mass transfer effects for a gas condensate field," in Proceedings of the SPE Asia Pacific Conference, Kuala Lumpur, Malaysia, March 1998, SPE 39746.

17. F. S. Alavi, D. Mowla, and F. Esmaeilzadeh, "Production performance analysis of Sarkhoon gas condensate reservoir," Journal of Petroleum Science and Engineering, vol. 75, no. 1-2, pp. 44–53, 2010. · ·

18. E. L. Zhu, Z. Wei, and Z. W. Wang, Natural Gas Extraction Process, Petroleum Industry Press, Beijing, China, 1993.
19. P. Y. Yang, G. X. Qi, and L. R. Chen, Condensate Gas Field Development, Petroleum Industry Press, Beijing, China, 1983.
20. D. Y. Peng and D. B. Robinson, "A new two-constant equation of state," Industrial and Engineering Chemistry: Fundamentals, vol. 15, pp. 59–64, 1976.
21. SY/T5542-2009, "Analysis for reservoir fluids physical properties," China National Oil and Gas Industry Standards, 2009.
22. A. H. Mohammadi, A. Chapoy, B. Tohidl, and D. Richon, "Gas solubility: a key to estimating the water content of natural gases," Industrial and Engineering Chemistry Research, vol. 45, no. 13, pp. 4825–4829, 2006. · ·
23. L. John and R. A. Walter, Gas Reservoir Engineering, Petroleum Industry Press, Beijing, 2007, Translated by Wang Y. P., Guo W. K., Pang Y. M. et al.

Chapter 6

Experimental and Modeling Studies on the Prediction of Gas Hydrate Formation

Jian-Yi Liu[1], Jing Zhang[2,3], Yan-Li Liu[4], Xiao-Hua Tan[1], and Jie Zhang[5]

[1]State Key Laboratory of Oil and Gas Reservoir Geology and Exploitation, Southwest Petroleum University, Xindu Road 8, Chengdu 610500, China

[2]Sulige Gas Field Research Center, Changqing Oilfield Company, Xi'an 710018, China

[3]National Engineering Laboratory for Exploration and Development of Low Permeability Oil and Gas Fields, Xi'an 710018, China

[4]Drilling and Producing Technology Research Institute, Liaohe Oilfield, Panjin 124010, China

[5]Tahe Oil Production Plant No.1, Sinopec Northwest Oilfield Branch, Luntai 841600, China

ABSTRACT

On the base of some kinetics model analysis and kinetic observation of hydrate formation process, a new prediction model of gas hydrate formation is proposed. The analysis of the present model shows that the formation of gas hydrate not only relevant with gas composition and free water content but also relevant with temperature and pressure. Through contrast experiment, the predicted result of the new prediction method of gas hydrate crystallization kinetics is close to measured result, it means that the prediction method can reflect the hydrate crystallization accurately.

INTRODUCTION

During the exploitation process, the gas well sometimes closed because of plugging in wellbore or surface pipeline which is caused by gas hydrate [1–4]. In order to solve a series of plugging problems caused by hydrate in oil field and develop natural gas hydrate resources more effectively and utilize hydrate technology more reasonably, the research on natural gas hydrate crystallization kinetics mechanism becomes very necessary [5,6].

An operation step of crystal nucleus growth process is an important factor need to research in the study of crystallization kinetics [7]. In different system, many factors can be used to control crystal nucleus growth process, such as diffusion, heat transfer, stirring rate, reaction kinetics, and heat exchange rate on surface of crystal [8–10]. In view of the above factors, many researchers propose the crystallization kinetics model which describes the nucleation based on mass transfer theory, crystallization theory, and two-film gas-liquid mass transfer theory [11, 12].

The generation of hydrate is actually the nucleation and crystal growth process. Kinetics of hydrate formation is related to generation rate, pressure, temperature, and so forth. The process can be clearly divided into three steps: at first the generation of crystal nucleus with critical radius, then solid crystal nucleus growth, and at last the components transfer to solid-liquid interface of nuclear at aggregation state [13–15].

Many factors affect the generation of natural gas hydrate; there are three main points. (1) When the gas temperature is equal to or lower than the water dew point and is accompanied with free water or liquid water, then hydrate formed. (2) In condition of certain pressure and gas composition, the temperature is lower than the hydrate formation temperature; then hydrate formed. (3) The high operating pressure will rise the hydrate formation temperature [16].

This paper developed a new prediction method of gas hydrate crystallization kinetics according to the shrinkage bubble model and considering the variation of pressure with time at the same time, to establish a new method for the prediction of natural gas hydrate.

MODEL

Hydrate Growth Dynamic

During the hydrate formation process there is transfer phenomena in the hydrate-gas-water [17–19]. The kinetic observation of hydrate formation process is shown in Figure 1. From the figure we can know that at first hydrate mainly gathered at the air-water interface layer; when the temperature is lower than the hydrate phase equilibrium temperature, hydrate nucleation formed quickly by gas and water molecules inside the bubble [20,21]. The generation process of crystal nucleus is the process that water diffuse into a bubble and gas in bubble diffuse out of bubble; finally the bubble in hydrate gradually reduced and even disappeared then hydrate crystals formed. Experiments show that the hydrate formation process is controlled by the transmission rate of gas and water molecules, through the establishment of hydrate formation kinetics method can describe the generation process [22].

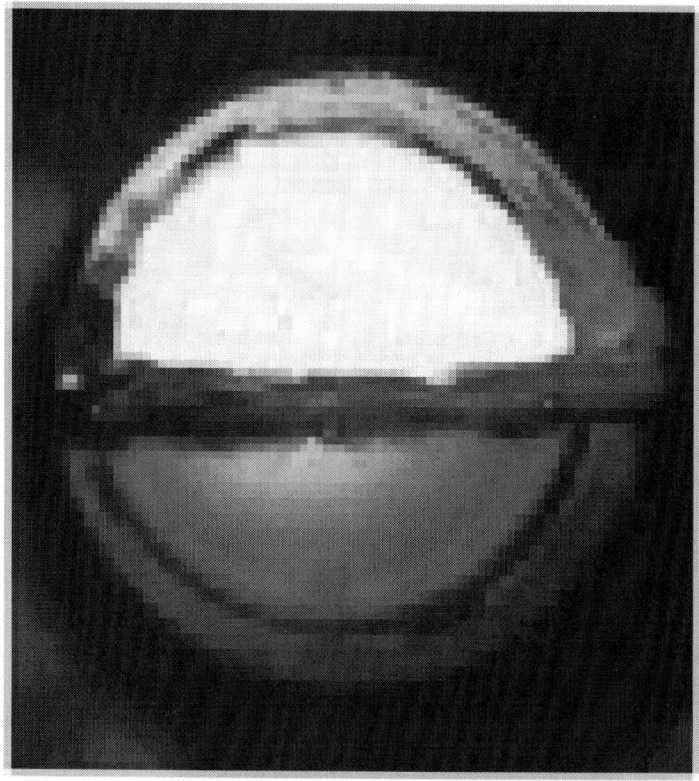

Figure 1: Kinetic observation of hydrate formation process.

Shrinkage Bubble Kinetics

Through the hydrate growth kinetic observation we can know that during the growth process of hydrate crystals the bubbles transform into hydrate; the process is shown in Figure 2. In Figure 2, the exterior is water, the internal is gas, and the ring is hydrate.

Experimental and Modeling Studies on the Prediction of Gas Hydrate...

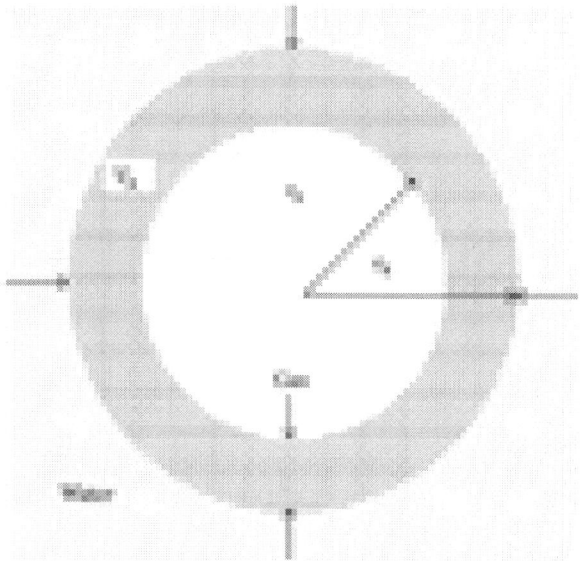

Figure 2: Process of a bubble transforming into hydrate.

Hydrate volume in hydrate crystals is

$$V_h = \frac{4}{3}\pi\left(r_h^3 - r_b^3\right), \tag{1}$$

where r_h is outer radius of hydrate, r_b is bubble radius, and V_h is hydrate volume in hydrate crystals.

The growth rate of hydrate is

$$\frac{dV_h}{dt} = 4\pi\left[r_h^2\frac{dr_h}{dt} - r_b^2\frac{dr_b}{dt}\right] \tag{2}$$

During the formation process of hydrates, the chemical potential in phase equilibrium systemization is equal. According to gas adsorption model and the law of diffusion, the hydrate formation rate equation is obtained as follows

$$\frac{dV_h}{dt} = 4\pi D \frac{r_b r_h}{r_h - r_b} \left[\frac{p_b - p_d}{p_b} \right] \tag{3}$$

where p_b is inner air pressure of bubble and p_d is decomposition pressure. D is the mass transfer coefficient and it can be expressed as

$$D = \frac{(1 - y_d) D_{ha}}{y_d} + vc_d D_{hw}, \tag{4}$$

where y_d is lattice occupancy, V is vacant hole numbers of each water molecule, c_d is water solution concentration under the decomposition pressure of hydrate, D_{ha} is mass transfer coefficient between hydrate and gas, and D_{hw} is mass transfer coefficient between hydrate and water

Supposing the gas in bubble and hydrate is ideal gas, then

$$V_b p_b + V_h p_{ha} \phi = V_{b0} p_{b0}, \tag{5}$$

where ϕ is porosity and V_b is bubble volume. p_{ha} is the pressure when the gas density in bubble is equal to the gas density in hydrate, $\rho_{ha'}$ which can be expressed as

$$p_{ha} = \frac{\rho_{ha} R_g T}{M}. \tag{6}$$

Because the reaction is controlled by mass transfer, so the change rate of hydrate volume is

$$\omega_h = -\frac{\left[dV_b/dt + (1 - d_{ha})(dV_b/dt)\right]}{(V_b + V_h)} \tag{7}$$

From the experiment we can know that hydrate is plastic materials, so

$$p_1 - p_b = \frac{4}{3}K_c(\omega - c\omega),$$
(8)

Where ω is compressibility of vacant hole.
Then forward formula becomes

$$p_1 - p_b = \frac{4}{3}K_c\omega_h\left[\frac{r_h^3 - r_b^3}{r_h^3}\right],$$

$$\frac{dV_b}{dt} = -\left[\frac{V_b}{p_b}\frac{dp_b}{dt} + \frac{p_{ha}\phi}{p_b}\frac{dV_h}{dt}\right].$$
(9)

By comprehensive analysis

$$r_h = \left[\frac{3(V_{b0}p_{b0} - V_b p_b)}{4\pi p_{ha}\phi} + r_b^3\right]^{1/3}$$
(10)

Dividing the formation process of gas hydrate into many small blocks of time and assuming that the pressure in bubbles is unchanged mean that p_b is a constant. From the above formulas the relationship that r_b changes with time can be known; then the rate of hydrate formation can be known, finally:

$$p_1 - p_b = \frac{4}{3}K_c\frac{r_h^3 - r_b^3}{r_h^3}$$

$$\times \left(p_b\left[\frac{(p_{ha}\phi/p_b - 1 + d_{ha})4\pi Dr_b r_h}{(r_h - r_b)(p_b - p_a)/p_h}\right]\right)$$

$$\times (V_h p_b + V_{b0}p_{b0} - V_h p_{ha}\phi)^{-1}.$$
(11)

From the above formulas r_h and r_b can be calculated. r_h and r_b are corresponding to p_b, and p_b is corresponding to t, so the relationship between $t-p_b$, r_h, and r_b at the whole reaction time can be calculated.

Through shrinkage bubble kinetics model the relationship between inner air pressure of bubble, outer radius of hydrate, bubble radius, and time can be known. But in the actual conditions, only environmental pressure can be detected, so it is necessary to consider relationship between environmental pressure and time to improve the shrinkage bubble kinetics.

Improvement of Shrinkage Bubble Kinetics Model

According to the constant volume substance equilibrium rule the alternation law of pressure with time can be known. In the experiments, the volume is constant which is equal to the volume of natural gas and water before hydrate formation and equal to the volume of residual gas, water, and hydrate after hydrate formation; then the pressure change. The process is shown in Figure 3.

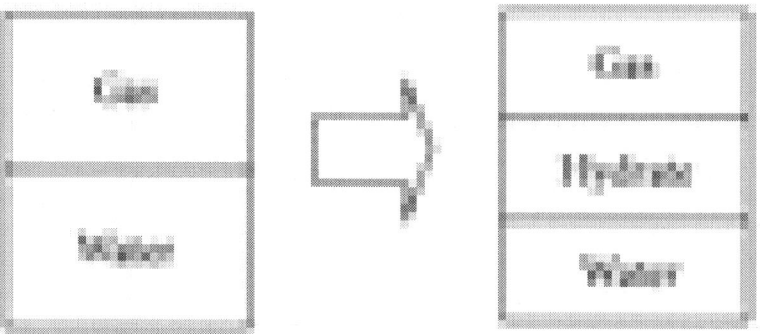

Figure 3: Chart of hydrate formation.

According to the principle of material balance, before hydrate formation there are

$$V_{sum} = V_g + V_w,$$

$$V_g = \frac{ZTp_{sc}V_{gsc}}{p_1 T_{sc}},$$

$$V_w = B_w V_{wsc}, \qquad (12)$$

where Vg is gas volume in the vessel, m³. V_w is water volume in the vessel, m³. Z is Z-factor at the vessel pressure, dimensionless. T is temperature in vessel; K. p_1 is the pressure in vessel before hydrate formation, MPa. T_{sc} is the temperature at standard state, K. p_{sc} is the pressure at standard state, MPa. V_{gsc} is the injected gas volume at standard state, m³. V_{wsc} is the injected water volume at standard state, m³. B_w is the volume coefficient of water, dimensionless.

After hydrate formation there are

$$V_{sum} = V'_g + V'_w + V_h,$$

$$V'_w = V_w - V_{wc},$$

$$p_2 = \frac{N_h RZT}{V_{sum} - V'_w - V_h},$$

$$V_{wc} = \frac{KN_h}{M_w \rho_w},$$

$$N_h = \frac{V_h \rho_h}{M_h},$$

$$V_h = nv_h,$$

$$n \leq \frac{S_{GW}}{\pi (3v_h/4\pi)^{2/3}}, \qquad (13)$$

Where V_g is gas volume in the vessel, m³. V_w is water volume in the vessel, m³. Vh is hydrate volume in the vessel, m³. V_{wc} is consumption water volume in the vessel, m³. p_2 is the pressure in the vessel after hydrate formation, MPa. N_h is molar of generated hydrate, K is molar of water when generate 1 mole hydrate. M_w is hydrate molecular weight, g/mol. ρ is hydrate density, kg/m^3. n is number of generated bubbles. V_h is hydrate volume of each generated bubbles, m^3. S_{GW} is interface area between gas and water, m^2.

EXPERIMENT

Gas Sample

In this experiment using the gas sample, the content of main gas components is measured by gas chromatography, and the components are shown in Table 1.

Table 1: Gas components

Components	N2	CO2	C1	C2	C3	iC4	nC4	iC5	nC5	C6
Content (mol %)	2.012	0.2659	82.9622	9.6578	3.5692	0.3044	0.6639	0.1687	0.2064	0.1894

Water Sample

In this experiment using the field water, the salinity of formation water is 14477 mg/L and pH value is 5.4. The analysis data of water is shown in Table 2.

Table 2: Water analysis

Item	K+	Na+	Ca2+	Mg2+	Cl−	SO42−	HCO3−	Total salinity
Content (mg/L)	365	4602	412	96	8007	535	460	14477

Injecting 10 mL formation water into vessel, then injecting gas until the pressure of vessel up to 14.7 MPa, and keeping the temperature at 20°C. It has been known that the temperature of hydrate formation is 18°C when the pressure is 14.7 MPa. By decreasing the temperature to 9°C we can obtain the supercooled degree, that is, 9°C. We observe and record the change of the pressure in the vessel.

THE RESULTS AND DISCUSSIONS

As shown in Figure 4, at first the pressure in the vessel declined rapidly; then the decrease rate become slow and steady gradually. The predicted pressure is close to measured pressure. But in the actual conditions, only environmental pressure can be detected, so it is just considered relationship between environmental pressure and time to improve the shrinkage bubble kinetics.

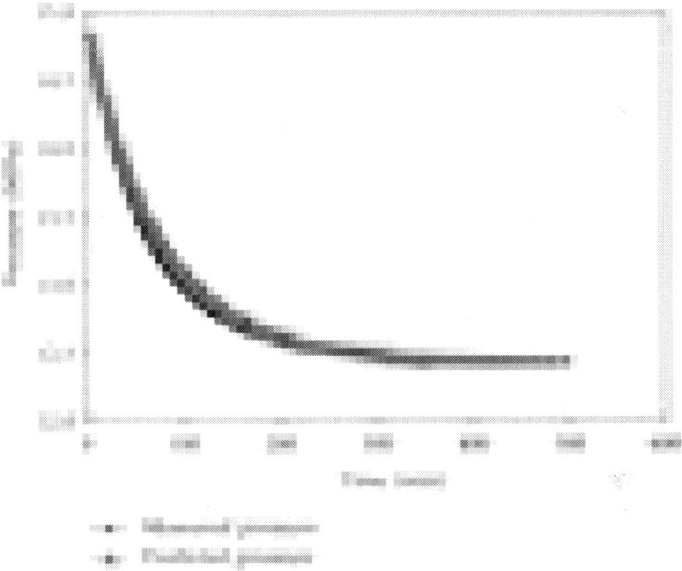

Figure 4: Comparison of the results of prediction method with experiment.

CONCLUSIONS

The predicted result of the new prediction method of gas hydrate crystallization kinetics is close to measured result; it means that the prediction method can reflect the hydrate crystallization accurately. In actual application the various parameters should be modified according to the practical situation.

ACKNOWLEDGMENTS

This work was jointly supported by the National Natural Science Foundation of China (51474181) and the 2014 Australia China Natural Gas Technology Partnership Fund Top Up Scholarships.

REFERENCES

1. E. D. Sloan, Hydrate Engineering, Society of Petroleum Engineers, Richardson, Tex, USA, 2000.
2. I. U. R. F. Makogon, Hydrates of Hydrocarbons, Pennwell Books, 1997.
3. X.-H. Tan, X.-P. Li, and J.-Y. Liu, "Model of continuous liquid removal from gas wells by droplet diameter estimation," Journal of Natural Gas Science and Engineering, vol. 15, pp. 8–13, 2013.
4. X.-H. Tan, J.-Y. Liu, X.-P. Li, G.-D. Zhang, and C. Tang, "A fractal model for the maximum droplet diameter in gas-liquid mist flow," Mathematical Problems in Engineering, vol. 2013, Article ID 532638, 6 pages, 2013.
5. J. Cai, E. Perfect, C.-L. Cheng, and X. Hu, "Generalized modeling of spontaneous imbibition based on hagen-poiseuille flow in tortuous capillaries with variably shaped apertures," Langmuir, vol. 30, no. 18, pp. 5142–5151, 2014.
6. X.-H. Tan, X.-P. Li, J.-Y. Liu, G.-D. Zhang, and L.-H. Zhang, "Analysis of permeability for transient two-phase flow in fractal porous media," Journal of Applied Physics, vol. 115, no. 11, Article ID 113502, 2014.
7. E. D. Sloan Jr. and C. Koh, Clathrate Hydrates of Natural Gases, CRC Press, New York, NY, USA, 2007.
8. W. R. Parrish and J. M. Prausnitz, "Dissociation pressures of gas hydrates formed by gas mixtures," Industrial & Engineering Chemistry Process Design and Development, vol. 11, no. 1, pp. 26–35, 1972.
9. H. Tavasoli, F. Feyzi, M. R. Dehghani, and F. Alavi, "Prediction of gas hydrate formation condition in the presence of thermodynamic inhibitors with the Elliott-Suresh-Donohue Equation of State,"

Journal of Petroleum Science and Engineering, vol. 77, no. 1, pp. 93–103, 2011.

10. H. Jiang and H. Adidharma, "Prediction of hydrate dissociation conditions for alkanes in the presence of alcohol and electrolyte solutions using ion-based statistical associating fluid theory," Chemical Engineering Science, vol. 82, pp. 14–21, 2012.

11. X.-S. Li, H.-J. Wu, and P. Englezos, "Prediction of gas hydrate formation conditions in the presence of methanol, glycerol, ethylene glycol, and triethylene glycol with the statistical associating fluid theory equation of state," Industrial & Engineering Chemistry Research, vol. 45, no. 6, pp. 2131–2137, 2006.

12. S. Moradi, A. Haghtalab, and A. Fazlali, "Prediction of hydrate formation conditions in the solutions containing electrolyte and alcohol inhibitors and their mixtures using UNIQUAC-NRF models," Fluid Phase Equilibria, vol. 349, pp. 61–66, 2013.

13. H.-J. Ng and D. B. Robinson, "The measurement and prediction of hydrate formation in liquid hydrocarbon-water systems," Industrial and Engineering Chemistry Fundamentals, vol. 15, no. 4, pp. 293–298, 1976.

14. N. Gnanendran and R. Amin, "Modelling hydrate formation kinetics of a hydrate promoter-water-natural gas system in a semi-batch spray reactor," Chemical Engineering Science, vol. 59, no. 18, pp. 3849–3863, 2004.

15. Y. Du and T.-M. Guo, "Prediction of hydrate formation for systems containing methanol," Chemical Engineering Science, vol. 45, no. 4, pp. 893–900, 1990.

16. T.-M. Guo, B.-H. Wu, Y.-H. Zhu, S.-S. Fan, and G.-J. Chen, "A review on the gas hydrate research in China," Journal of Petroleum Science and Engineering, vol. 41, no. 1–3, pp. 11–20, 2004.

17. C. A. Koh, E. D. Sloan, A. K. Sum, and D. T. Wu, "Fundamentals and applications of gas hydrates," Annual Review of Chemical and Biomolecular Engineering, vol. 2, pp. 237–257, 2011.

18. J. Javanmardi, S. Babaee, A. Eslamimanesh, and A. H. Mohammadi, "Experimental measurements and predictions of gas hydrate dissociation conditions in the presence of methanol and ethane-1,2-diol aqueous solutions," Journal of Chemical and Engineering Data, vol. 57, no. 5, pp. 1474–1479, 2012.

19. Y. F. Makogon and R. Y. Omelchenko, "Commercial gas production from Messoyakha deposit in hydrate conditions," Journal of Natural Gas Science and Engineering, vol. 11, pp. 1–6, 2013.
20. M. Mottahedin, F. Varaminian, and K. Mafakheri, "Modeling of methane and ethane hydrate formation kinetics based on non-equilibrium thermodynamics," Journal of Non-Equilibrium Thermodynamics, vol. 36, no. 1, pp. 3–22, 2011.
21. K. Nasrifar and M. Moshfeghian, "A model for prediction of gas hydrate formation conditions in aqueous solutions containing electrolytes and/or alcohol," The Journal of Chemical Thermodynamics, vol. 33, no. 9, pp. 999–1014, 2001.
22. A. Eslamimanesh, A. H. Mohammadi, and D. Richon, "Thermodynamic consistency test for experimental solubility data in carbon dioxide/methane + water system inside and outside gas hydrate formation region," Journal of Chemical & Engineering Data, vol. 56, no. 4, pp. 1573–1586, 2011.

Microbial Degradation of Petroleum Hydrocarbon Contaminants: An Overview

Nilanjana Das and Preethy Chandran

Environmental Biotechnology Division, School of Biosciences and Technology, VIT University, Vellore, Tamil Nadu 632014, India

ABSTRACT

One of the major environmental problems today is hydrocarbon contamination resulting from the activities related to the petrochemical industry. Accidental releases of petroleum products are of particular concern in the environment. Hydrocarbon components have been known to belong to the family of carcinogens and neurotoxic organic pollutants. Currently accepted disposal methods of incineration or burial insecure landfills can become prohibitively expensive when amounts of contaminants are large. Mechanical and chemical methods generally used to remove hydrocarbons from contaminated sites have limited effectiveness and can be expensive. Bioremediation is the promising technology for the treatment of these contaminated sites since it is cost-effective and will lead to complete mineralization. Bioremediation functions basically on biodegradation, which may refer to complete mineralization of organic contaminants into carbon dioxide, water, inorganic compounds, and cell protein or transformation of complex organic contaminants to other simpler organic compounds by biological agents like microorganisms. Many indigenous microorganisms in water and soil are capable of degrading hydrocarbon contaminants. This paper presents an updated overview of petroleum hydrocarbon degradation by microorganisms under different ecosystems.

INTRODUCTION

Petroleum-based products are the major source of energy for industry and daily life. Leaks and accidental spills occur regularly during the exploration, production, refining, transport, and storage of petroleum and petroleum products. The amount of natural crude oil seepage was estimated to be 600,000 metric tons per year with a range of uncertainty of 200,000 metric tons per year [1]. Release of hydrocarbons into the environment whether accidentally or due to human activities is a main cause of water and soil pollution [2]. Soil contamination with hydrocarbons causes extensive damage of local system since accumulation of pollutants in animals and plant tissue may cause death or mutations [3]. The technology commonly used for the soil remediation includes mechanical, burying, evaporation, dispersion, and washing. However, these technologies are expensive and can lead to incomplete decomposition of contaminants.

The process of bioremediation, defined as the use of microorganisms to detoxify or remove pollutants owing to their diverse metabolic capabilities is an evolving method for the removal and degradation of many environmental pollutants including the products of petroleum industry [4]. In addition, bioremediation technology is believed to be noninvasive and relatively cost-effective [5]. Biodegradation by natural populations of microorganisms represents one of the primary mechanisms by which petroleum and other hydrocarbon pollutants can be removed from the environment [6] and is cheaper than other remediation technologies [7].

The success of oil spill bioremediation depends on one's ability to establish and maintain conditions that favor enhanced oil biodegradation rates in the contaminated environment. Numerous scientific review articles have covered various factors that influence the rate of oil biodegradation [7–12]. One important requirement is the presence of microorganisms with the appropriate metabolic capabilities. If these microorganisms are present, then optimal rates of growth and hydrocarbon biodegradation can be sustained by ensuring that adequate concentrations of nutrients and oxygen are present and that the pH is between 6 and 9. The physical and chemical characteristics of the oil and oil surface area are also important determinants of bioremediation success. There are the two main approaches to oil spill bioremediation: (a) bioaugmentation, in which known oil-degrading bacteria are added to supplement the existing microbial population, and (b) biostimulation, in which the growth of indigenous oil degraders is stimulated by the addition of nutrients or other growth-limiting cosubstrates.

The success of bioremediation efforts in the cleanup of the oil tanker Exxon Valdez oil spill of 1989 [13] in Prince William Sound and the Gulf of Alaska created tremendous interest in the potential of biodegradation and bioremediation technology. Most existing studies have concentrated on evaluating the factors affecting oil bioremediation or testing favored products and methods through laboratory studies [14]. Only limited numbers of pilot scale and field trials have provided the most convincing demonstrations of this technology which have been reported in the peer-reviewed literature [15–18]. The scope of current understanding of oil bioremediation is also limited because the emphasis of most of these field studies and reviews has been given on the evaluation of bioremediation technology for dealing with large-scale oil spills on marine shorelines.

This paper provides an updated information on microbial degradation of petroleum hydrocarbon contaminants towards the better understanding in bioremediation challenges.

MICROBIAL DEGRADATION OF PETROLEUM HYDROCARBONS

Biodegradation of petroleum hydrocarbons is a complex process that depends on the nature and on the amount of the hydrocarbons present. Petroleum hydrocarbons can be divided into four classes: the saturates, the aromatics, the asphaltenes (phenols, fatty acids, ketones, esters, and porphyrins), and the resins (pyridines, quinolines, carbazoles, sulfoxides, and amides) [19]. Different factors influencing hydrocarbon degradation have been reported by Cooney et al. [20]. One of the important factors that limit biodegradation of oil pollutants in the environment is their limited availability to microorganisms. Petroleum hydrocarbon compounds bind to soil components, and they are difficult to be removed or degraded [21]. Hydrocarbons differ in their susceptibility to microbial attack. The susceptibility of hydrocarbons to microbial degradation can be generally ranked as follows: linear alkanes > branched alkanes > small aromatics > cyclic alkanes [6, 22]. Some compounds, such as the high molecular weight polycyclic aromatic hydrocarbons (PAHs), may not be degraded at all [23].

Microbial degradation is the major and ultimate natural mechanism by which one can cleanup the petroleum hydrocarbon pollutants from the environment [24–26]. The recognition of biodegraded petroleum-derived aromatic hydrocarbons in marine sediments was reported by Jones et al. [27]. They studied the extensive biodegradation of alkyl aromatics in marine sediments which occurred prior to detectable biodegradation of n-alkane profile of the crude oil and the microorganisms, namely, Arthrobacter, Burkholderia, Mycobacterium, Pseudomonas, Sphingomonas, and Rhodococcus were found to be involved for alkylaromatic degradation. Microbial degradation of petroleum hydrocarbons in a polluted tropical stream in Lagos, Nigeria was reported by Adebusoye et al. [28]. Nine bacterial strains, namely, Pseudomonas fluorescens, P. aeruginosa, Bacillus subtilis, Bacillus sp., Alcaligenes sp., Acinetobacter lwoffi, Flavobacterium sp., Micrococcus roseus, and Corynebacterium sp. were isolated from the polluted stream which could degrade crude oil.

Hydrocarbons in the environment are biodegraded primarily by bacteria, yeast, and fungi. The reported efficiency of biodegradation ranged from 6% [29] to 82% [30] for soil fungi, 0.13% [29] to 50%

[30] for soil bacteria, and 0.003% [31] to 100% [32] for marine bacteria. Many scientists reported that mixed populations with overall broad enzymatic capacities are required to degrade complex mixtures of hydrocarbons such as crude oil in soil [33], fresh water [34], and marine environments [35, 36].

Bacteria are the most active agents in petroleum degradation, and they work as primary degraders of spilled oil in environment [37, 38]. Several bacteria are even known to feed exclusively on hydrocarbons [39]. Floodgate [36] listed 25 genera of hydrocarbon degrading bacteria and 25 genera of hydrocarbon degrading fungi which were isolated from marine environment. A similar compilation by Bartha and Bossert [33] included 22 genera of bacteria and 31 genera of fungi. In earlier days, the extent to which bacteria, yeast, and filamentous fungi participate in the biodegradation of petroleum hydrocarbons was the subject of limited study, but appeared to be a function of the ecosystem and local environmental conditions [7]. Crude petroleum oil from petroleum contaminated soil from North East India was reported by Das and Mukherjee [40]. Acinetobacter sp. was found to be capable of utilizing n-alkanes of chain length C_{10}–C_{40} as a sole source of carbon [41]. Bacterial genera, namely, Gordonia, Brevibacterium, Aeromicrobium, Dietzia, Burkholderia, and Mycobacterium isolated from petroleum contaminated soil proved to be the potential organisms for hydrocarbon degradation [42]. The degradation of poly-aromatic hydrocarbons by Sphingomonas was reported by Daugulis and McCracken [43].

Fungal genera, namely, Amorphoteca, Neosartorya, Talaromyces, and Graphium and yeast genera, namely, Candida, Yarrowia, and Pichia were isolated from petroleum-contaminated soil and proved to be the potential organisms for hydrocarbon degradation [42]. Singh [44] also reported a group of terrestrial fungi, namely, Aspergillus, Cephalosporium, and Pencillium which were also found to be the potential degrader of crude oil hydrocarbons. The yeast species, namely, Candida lipolytica, Rhodotorula mucilaginosa, Geotrichum sp, and Trichosporon mucoides isolated from contaminated water were noted to degrade petroleum compounds [45].

Though algae and protozoa are the important members of the microbial community in both aquatic and terrestrial ecosystems, reports are scanty regarding their involvement in hydrocarbon biodegradation. Walker et al. [51] isolated an alga, Prototheca zopfi which was capable of utilizing crude oil and a mixed hydrocarbon substrate and exhibited extensive degradation of n-alkanes and isoalkanes as well as aromatic hydrocarbons. Cerniglia et al. [52] observed that nine cyanobacteria, five green algae, one red alga, one brown alga, and two diatoms could oxidize naphthalene. Protozoa, by contrast, had not been shown to utilize hydrocarbons.

FACTORS INFLUENCING PETROLEUM HYDROCARBON DEGRADATION

A number of limiting factors have been recognized to affect the biodegradation of petroleum hydrocarbons, many of which have been discussed by Brusseau [53]. The composition and inherent biodegradability of the petroleum hydrocarbon pollutant is the first and foremost important consideration when the suitability of a remediation approach is to be assessed. Among physical factors, temperature plays an important role in biodegradation of hydrocarbons by directly affecting the chemistry of the pollutants as well as affecting the physiology and diversity of the microbial flora. Atlas [54] found that at low temperatures, the viscosity of the oil increased, while the volatility of the toxic low molecular weight hydrocarbons were reduced, delaying the onset of biodegradation.

Temperature also affects the solubility of hydrocarbons [62]. Although hydrocarbon biodegradation can occur over a wide range of temperatures, the rate of biodegradation generally decreases with the decreasing temperature. Figure 1 shows that highest degradation rates that generally occur in the range 30–40°C in soil environments, 20–30°C in some freshwater environments and 15–20°C in marine environments [33, 34]. Venosa and Zhu [63] reported that ambient temperature of the environment affected both the properties of spilled oil and the activity of the microorganisms. Significant biodegradation of hydrocarbons have been reported in psychrophilic environments in temperate regions [64, 65].

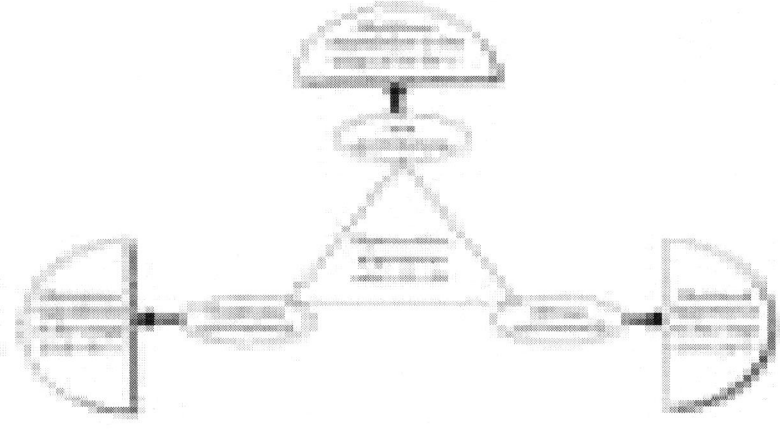

Figure 1: Hydrocarbon degradation rates in soil, fresh water, and marine environments.

Nutrients are very important ingredients for successful biodegradation of hydrocarbon pollutants especially nitrogen, phosphorus, and in some cases iron [34]. Some of these nutrients could become limiting factor thus affecting the biodegradation processes. Atlas [35] reported that when a major oil spill occurred in marine and freshwater environments, the supply of carbon was significantly increased and the availability of nitrogen and phosphorus generally became the limiting factor for oil degradation. In marine environments, it was found to be more pronounced due to low levels of nitrogen and phosphorous in seawater [36]. Freshwater wetlands are typically considered to be nutrient deficient due to heavy demands of nutrients by the plants [66]. Therefore, additions of nutrients were necessary to enhance the biodegradation of oil pollutant [67, 68]. On the other hand, excessive nutrient concentrations can also inhibit the biodegradation activity [69]. Several authors have reported the negative effects of high NPK levels on the biodegradation of hydrocarbons [70, 71] especially on aromatics [72]. The effectiveness of fertilizers for the crude oil bioremediation in subarctic intertidal sediments was studied by Pelletier et al. [64]. Use of poultry manure as organic fertilizer in contaminated soil was also reported [73], and biodegradation was found to be enhanced in the presence of poultry manure alone. Maki et al. [74] reported that photo-oxidation increased the biodegradability of petroleum hydrocarbon by increasing its bioavailability and thus enhancing microbial activities.

MECHANISM OF PETROLEUM HYDROCARBON DEGRADATION

The most rapid and complete degradation of the majority of organic pollutants is brought about under aerobic conditions. Figure 2 shows the main principle of aerobic degradation of hydrocarbons [75]. The initial intracellular attack of organic pollutants is an oxidative process and the activation as well as incorporation of oxygen is the enzymatic key reaction catalyzed by oxygenases and peroxidases. Peripheral degradation pathways convert organic pollutants step by step into intermediates of the central intermediary metabolism, for example, the tricarboxylic acid cycle. Biosynthesis of cell biomass occurs from the central precursor metabolites, for example, acetyl-CoA, succinate, pyruvate. Sugars required for various biosyntheses and growth are synthesized by gluconeogenesis.

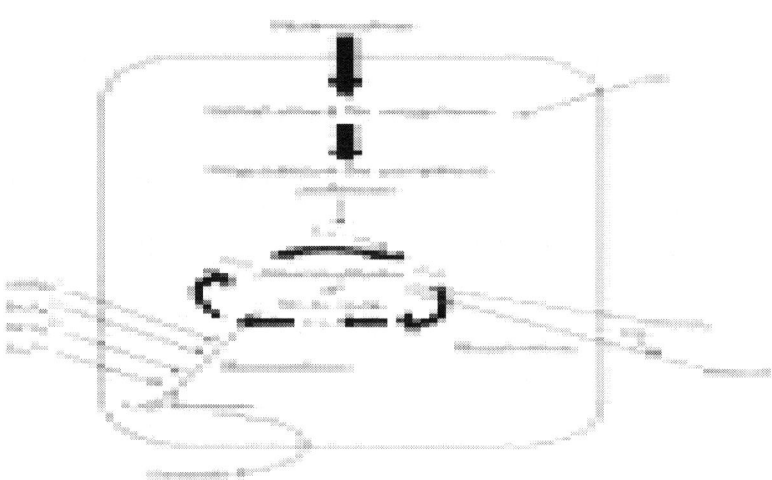

Figure 2: Main principle of aerobic degradation of hydrocarbons by microorganisms.

The degradation of petroleum hydrocarbons can be mediated by specific enzyme system. Figure 3 shows the initial attack on xenobiotics by oxygenases [75]. Other mechanisms involved are (1) attachment of microbial cells to the substrates and (2) production of biosurfactants

[76]. The uptake mechanism linked to the attachment of cell to oil droplet is still unknown but production of biosurfactants has been well studied.

Figure 3: Enzymatic reactions involved in the processes of hydrocarbons degradation.

ENZYMES PARTICIPATING IN DEGRADATION OF HYDROCARBONS

Cytochrome P450 alkane hydroxylases constitute a super family of ubiquitous Heme-thiolate Monooxygenases which play an important role in the microbial degradation of oil, chlorinated hydrocarbons, fuel additives, and many other compounds [77]. Depending on the chain length, enzyme systems are required to introduce oxygen in the substrate to initiate biodegradation (Table 1). Higher eukaryotes generally contain several different P450 families that consist of large number of individual P450 forms that may contribute as an ensemble of isoforms to the metabolic conversion of given substrate. In microorganisms such P450 multiplicity can only be found in few species [78]. Cytochrome P450 enzyme systems was found to be involved in biodegradation of petroleum hydrocarbons (Table 1). The capability of several yeast species to use n-alkanes and other aliphatic hydrocarbons as a sole source of carbon and energy is mediated by the existence of multiple microsomal Cytochrome P450 forms. These cytochrome P450 enzymes had been isolated from yeast species such as Candida maltosa, Candida tropicalis, and Candida apicola [79]. The diversity of alkaneoxygenase systems in prokaryotes and eukaryotes that are actively participating in the degradation of alkanes under aerobic conditions like Cytochrome P450 enzymes, integral membrane di-iron alkane hydroxylases (e.g., alkB), soluble di-iron methane monooxygenases, and membrane-bound copper containing methane monooxygenases have been discussed by Van Beilen and Funhoff [80].

Table 1: Enzymes involved in biodegradation of petroleum hydrocarbons

Enzymes	Substrates	Microorganisms	References
Soluble Methane Monooxygenases	C1–C8 alkanes alkenes and cycloalkanes	Methylococcus	
		Methylosinus	
		Methylocystis	McDonald et al. [46]
		Methylomonas	
		Methylocella	
Particulate Methane Monooxygenases	C1–C5 (halogenated) alkanes and cycloalkanes	Methylobacter	
		Methylococcus,	McDonald et al. [46]
		Methylocystis	
AlkB related Alkane Hydroxylases	C5–C16 alkanes, fatty acids, alkyl benzenes, cycloalkanes and so forth	Pseudomonas	
		Burkholderia	Jan et al. [47]
		Rhodococcus,	
		Mycobacterium	
Eukaryotic P450	C10–C16 alkanes, fatty acids	Candida maltosa	
		Candida tropicalis	Iida et al. [48]
		Yarrowia lipolytica	

Bacterial P450 oxygenase system	C5–C16 alkanes, cycloalkanes	Acinetobacter	Van Beilen et al. [49]
		Caulobacter	
		Mycobacterium	
Dioxygenases	C10–C30 alkanes	Acinetobacter sp.	Maeng et al. [50]

UPTAKE OF HYDROCARBONS BY BIOSURFACTANTS

Biosurfactants are heterogeneous group of surface active chemical compounds produced by a wide variety of microorganisms [57, 58, 60, 81–83]. Surfactants enhance solubilization and removal of contaminants [84, 85]. Biodegradation is also enhanced by surfactants due to increased bioavailability of pollutants [86]. Bioremediation of oil sludge using biosurfactants has been reported by Cameotra and Singh [87]. Microbial consortium consisting of two isolates of Pseudomonas aeruginosa and one isolate Rhodococcus erythropolisfrom soil contaminated with oily sludge was used in this study. The consortium was able to degrade 90% of hydrocarbons in 6 weeks in liquid culture. The ability of the consortium to degrade sludge hydrocarbons was tested in two separate field trials. In addition, the effect of two additives (a nutrient mixture and a crude biosurfactant preparation on the efficiency of the process was also assessed. The biosurfactant used was produced by a consortium member and was identified as being a mixture of 11 rhamnolipid congeners. The consortium degraded 91% of the hydrocarbon content of soil contaminated with 1% (v/v) crude oil sludge in 5 weeks. Separate use of any one additive along with the consortium brought about a 91–95% depletion of the hydrocarbon content in 4 weeks, with the crude biosurfactant preparation being a more effective enhancer of degradation. However, more than 98% hydrocarbon depletion was obtained when both additives were added together with the consortium. The data substantiated the use of a crude biosurfactant for hydrocarbon remediation.

Pseudomonads are the best known bacteria capable of utilizing hydrocarbons as carbon and energy sources and producing biosurfactants

[37, 87–89]. Among Pseudomonads, P. aeruginosa is widely studied for the production of glycolipid type biosurfactants. However, glycolipid type biosurfactants are also reported from some other species like P. putida and P. chlororaphis. Biosurfactants increase the oil surface area and that amount of oil is actually available for bacteria to utilize it [90]. Table 2 summarizes the recent reports on biosurfactant production by different microorganisms. Biosurfactants can act as emulsifying agents by decreasing the surface tension and forming micelles. The microdroplets encapsulated in the hydrophobic microbial cell surface are taken inside and degraded. Figure 4 demonstrates the involvement of biosurfactant (rhamnolipids) produced by Pseudomonas sp. and the mechanism of formation of micelles in the uptake of hydrocarbons [75].

Table 2: Biosurfactants produced by microorganisms

Biosurfactants	Microorganisms
Sophorolipids	Candida bombicola (Daverey and Pakshirajan, [55])
Rhamnolipids	Pseudomonas aeruginosa (Kumar et al. [56])
Lipomannan	Candida tropicalis (Muthuswamy et al. [57])
Rhamnolipids	Pseudomonas fluorescens (Mahmound et al. [58])
Surfactin	Bacillus subtilis (Youssef et al. [59])
Glycolipid	Aeromonas sp. (Ilori et al. [60])
Glycolipid	Bacillus sp. (Tabatabaee et al. [61])

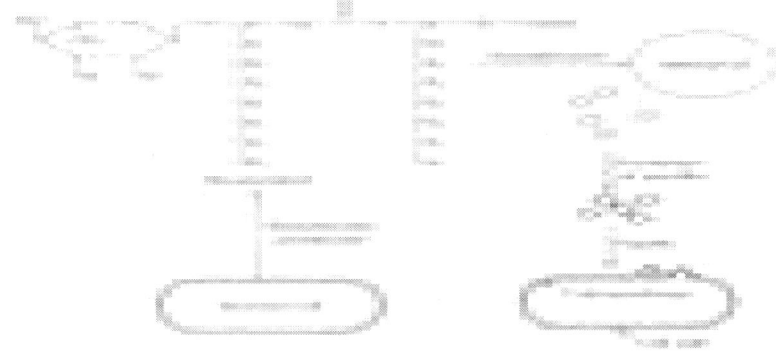

Figure 4: Involvement of biosurfactant (rhamnolipid) produced by Pseudomonas sp in the uptake of hydrocarbons.

BIODEGRADATION OF PETROLEUM HYDROCARBONS BY IMMOBILIZED CELLS

Immobilized cells have been used and studied for the bioremediation of numerous toxic chemicals. Immobilization not only simplifies separation and recovery of immobilized cells but also makes the application reusable which reduces the overall cost. Wilsey and Bradely [91] used free suspension and immobilizedPseudomonas sp. to degrade petrol in an aqueous system. The study indicated that immobilization resulted in a combination of increased contact between cell and hydrocarbon droplets and enhanced level of rhamnolipids production. Rhamnolipids caused greater dispersion of water-insoluble n-alkanes in the aqueous phase due to their amphipathic properties and the molecules consist of hydrophilic and hydrophobic moieties reduced the interfacial tension of oil-water systems. This resulted in higher interaction of cells with solubilized hydrocarbon droplets much smaller than the cells and rapid uptake of hydrocarbon in to the cells. Diaz et al. [92] reported that immobilization of bacterial cells enhanced the biodegradation rate of crude oil compared to free living cells in a wide range of culture salinity. Immobilization can be done in batch mode as well as continuous mode. Packed bed reactors are commonly used in continuous mode to degrade hydrocarbons. Cunningham et al. [93] used polyvinyl alcohol (PVA) cryogelation as an entrapment matrix and microorganisms indigenous to the site. They constructed laboratory biopiles to compare immobilised bioaugmentation with liquid culture bioaugmentation and biostimulation. Immobilised systems were found to be the most successful in terms of percentage removal of diesel after 32 days.

Rahman et al. [94] conducted an experiment to study the capacity of immobilized bacteria in alginate beads to degrade hydrocarbons. The results showed that there was no decline in the biodegradation activity of the microbial consortium on the repeated use. It was concluded that immobilization of cells are a promising application in the bioremediation of hydrocarbon contaminated site.

COMMERCIALLY AVAILABLE BIOREMEDIATION AGENTS

Microbiological cultures, enzyme additives, or nutrient additives that significantly increase the rate of biodegradation to mitigate the effects of the discharge were defied as bioremediation agents by U.S.EPA [95]. Bioremediation agents are classified as bioaugmentation agents and biostimulation agents based on the two main approaches to oil spill bioremediation. Numerous bioremediation products have been proposed and promoted by their vendors, especially during early 1990s, when bioremediation was popularized as "the ultimate solution" to oil spills [96].

The U.S. EPA compiled a list of 15 bioremediation agents [95, 97] as a part of the National Oil and Hazardous Substances Pollution Contingency Plan (NCP) Product Schedule, which was required by the Clean Water Act, the Oil Pollution Act of 1990, and the National Contingency Plan (NCP) as shown in Table 3. But the list was modified, and the number of bioremediation agents was reduced to nine.

Table 3: Bioremediation agents in NCP product schedule (Adapted from USE-PA, 2002)

Name or Trademark	Product Type	Manufacture
BET BIOPETRO	MC	BioEnviro Tech, Tomball, TX
BILGEPRO	NA	International Environmental Products, LLC, Conshohocken, PA.
INIPOL EAP 22	NA	Societe, CECA S.A., France
LAND AND SEA	NA	Land and Sea Restoration LLC, San Antonio, TX
RESTORATION MICRO-BLAZE	MC	Verde Environmental, Inc., Houston, TX
OIL SPILL EATER II	NA/EA	Oil Spill Eater International, Corporation, Dallas, TX
OPPENHEIMER FORMULA	MC	Oppenheimer Biotechnology, Inc., Austin, TX
PRISTINE SEA II	MC	Marine Systems, Baton Rouge, LA
STEP ONE	MC	B & S Research, Inc., Embarrass, MN
SYSTEM E.T. 20.	MC	Quantum Environmental Technologies, Inc(QET), La Jolla, CA
VB591TMWATER, VB997TMSOIL, AND BINUTRIX	NA	BioNutraTech, Inc., Houston, TX
WMI-2000	MC	WMI International, Inc

Abbreviations of product type:

MC: Microbial Culture

EA: Enzyme Additive

NA: Nutrient Additive.

Studies showed that bioremediation products may be effective in the laboratory but significantly less so in the field [14, 17, 18, 98]. This is because laboratory studies cannot always simulate complicated real world conditions such as spatial heterogeneity, biological interactions, climatic effects, and nutrient mass transport limitations. Therefore, field studies and applications are the ultimate tests or the most convincing demonstration of the effectiveness of bioremediation products.

Compared to microbial products, very few nutrient additives have been developed and marketed specifically as commercial bioremediation agents for oil spill cleanup. It is probably because common fertilizers are inexpensive, readily available, and have been shown effective if used properly. However, due to the limitations of common fertilizers (e.g., being rapidly washed out due to tide and wave action), several organic nutrient products, such as oleophilic nutrient products, have recently been evaluated and marketed as bioremediation agents. Four agents, namely, Inipol EAP22, Oil Spill Eater II (OSE II), BIOREN 1, and BIOREN 2, listed on the NCP Product Schedule have also been put into this category.

Inipol EAP22 (Societe, CECA S.A., France) is listed on the NCP Product Schedule as a nutrient additive and probably the most well-known bioremediation agent for oil spill cleanup due to its use in Prince William Sound, Alaska. This nutrient product is a microemulsion-containing urea as a nitrogen source, sodium laureth phosphate as a phosphorus source, 2-butoxy-1-ethanol as a surfactant, and oleic acid to give the material its hydrophobicity. The claimed advantages of Inipol EAP22 include (1) preventing the formation of water-in-oil emulsions by reducing the oil viscosity and interfacial tension; (2) providing controlled release of nitrogen and phosphorus for oil biodegradation; (3) exhibiting no toxicity to flora and fauna and good biodegradability [99].

Oil Spill Eater II (Oil Spill Eater International, Corp.) is another nutrient product listed on the NCP Schedule [97]. This product is listed as a nutrient/enzyme additive and consists of "nitrogen, phosphorus, readily available carbon, and vitamins for quick colonization of naturally occurring bacteria". A field demonstration was carried out at a bioventing site in a Marine Corps Air Ground Combat Center (MCAGCC) in California to investigate the efficacy of OSEII for enhancing hydrocarbon biodegradation in a fuel-contaminated vadose zone [106].

Researchers from European EUREKA BIOREN program conducted a field trial in an estuary environment to evaluate the effectiveness of two bioremediation products (BIOREN 1 and 2) [114, 115]. The two nutrient products were derived from fish meals in a granular form with urea and super phosphate as nitrogen and phosphorus sources and proteinaceous material as the carbon source. The major difference between the two formulations was that BIOREN 1 contained a biosurfactant. The results showed that the presence of biosurfactant in BIOREN 1 was the most active ingredient which contributed to the increase in oil degradation rates whereas BIOREN 2 (without biosurfactant) was not effective in that respect. The biosurfactant could have contributed to greater bioavailability of hydrocarbons to microbial attack.

PHYTOREMEDIATION

Phytoremediation is an emerging technology that uses plants to manage a wide variety of environmental pollution problems, including the cleanup of soils and groundwater contaminated with hydrocarbons and other hazardous substances. The different mechanisms, namely, hydraulic control, phytovolatilization, rhizoremediation, and phytotransformation. could be utilized for the remediation of a wide variety of contaminants.

Phytoremediation can be cost-effective (a) for large sites with shallow residual levels of contamination by organic, nutrient, or metal pollutants, where contamination does not pose an imminent danger and only "polishing treatment" is required; (b) where vegetation is used as a final cap and closure of the site [116].

Advantages of using phytoremediation include cost-effectiveness, aesthetic advantages, and long-term applicability (Table 4). Furthermore, the use of phytoremediation as a secondary or polishing in situ treatment step minimizes land disturbance and eliminates transportation and liability costs associated with offsite treatment and disposal.

Table 4: Advantages and disadvantages of phytoremediation over traditional technologies

Advantages	Disadvantages
Relatively low cost	Longer remediation times

Easily implemented and maintained	Climate dependent
Several mechanisms for removal	Effects to food web might be unknown
Environmentally friendly	Ultimate contaminant fates might be unknown
Aesthetically pleasing	Results are variable
Reduces landfilled wastes	
Harvestable plant material	

Research and application of phytoremediation for the treatment of petroleum hydrocarbon contamination over the past fifteen years have provided much useful information that can be used to design effective remediation systems and drive further improvement and innovation. Phytoremediation could be applied for the remediation of numerous contaminated sites. However, not much is known about contaminant fate and transformation pathways, including the identity of metabolites (Table 4). Little data exists on contaminant removal rates and efficiencies directly attributable to plants under field conditions.

The potential use of phytoremediation at a site contaminated with hydrocarbons was investigated. The Alabama Department of Environmental Management granted a site, which involved about 1500 cubic yards of soil of which 70% of the baseline samples contained over 100 ppm of total petroleum hydrocarbon (TPH). After 1 year of vegetative cover, approximately 83% of the samples were found to contain less than 10-ppm TPH. Removal of total petroleum hydrocarbon (TPH) at several field sites contaminated with crude oil, diesel fuel, or petroleum refinery wastes, at initial TPH concentrations of 1,700 to 16,000 mg/kg were also investigated [117, 118]. Plant growth was found to vary depending upon the species. Presence of some species led to greater TPH disappearance than with other species or in unvegetated soil. Among tropical plants tested for use in Pacific Islands, three coastal trees, kou (Cordia subcordata), milo (Thespesia populnea), and kiawe (Prosopis pallida) and the native shrub beach naupaka (*Scaevola serica*) tolerated field conditions and facilitated cleanup of soils contaminated with diesel fuel [119]. Grasses were often planted with trees at sites with organic contaminants as the primary remediation method. Tremendous amount of fine roots in the surface soil was found to be effective at binding and transforming hydrophobic contaminants such as TPH, BTEX, and PAHs. Grasses were often planted between rows of trees to provide soil stabilization and protection against wind-blown dust that could move contaminants offsite. Legumes such as alfalfa (Medicago sativa), alsike clover (Trifolium hybridum), and peas (Pisum sp.) could be used to restore nitrogen to poor soils. Fescue (Vulpia myuros), rye (Elymus sp.), clover (Trifolium sp.), and reed canary grass (Phalaris arundinacea) were used successfully at several sites, especially contaminated with petrochemical wastes. Once harvested, the grasses could be disposed off as compost or burned.

Microbial degradation in the rhizosphere might be the most significant mechanism for removal of diesel range organics in vegetated contaminated soils [120]. This occurs because contaminants such as PAHs are highly hydrophobic, and their sorption to soil decreases their bioavailability for plant uptake and phytotransformation.

GENETICALLY MODIFIED BACTERIA

Applications for genetically engineered microorganisms (GEMs) in bioremediation have received a great deal of attention to improve the degradation of hazardous wastes under laboratory conditions. There are reports on the degradation of environmental pollutants by different bacteria. Table 5 shows some examples of the relevant use of genetic engineering technology to improve bioremediation of hydrocarbon contaminants using bacteria. The genetically engineered bacteria showed higher degradative capacity. However, ecological and environmental concerns and regulatory constraints are major obstacles for testing GEM in the field. These problems must be solved before GEM can provide an effective clean-up process at lower cost.

Table 5: Genetic engineering for biodegradation of contaminants

Microorganisms	Modification	Contaminants	Reference
Pseudomonas. putida	pathway	4-ethylbenzoate	Ramos et al. [100]
P. putida KT2442	pathway	toluene/benzoate	Panke and Sanchezromero [101]

Pseudomonas sp.FRI	pathway	chloro-, methylbenzoates	Rojo et al. [102]
Comamonas. testosteroni VP44	substrate specificity	o-, p-monochlorobiphenyls	Hrywna et al. [103]
Pseudomonas sp. LB400	substrate specificity	PCB	Erickson and Mondello [104]
P. pseudoalcaligenes KF707-D2	substrate specificity	TCE, toluene, benzene	Suyama et al. [105]

The use of genetically engineered bacteria was applied to bioremediation process monitoring, strain monitoring, stress response, end-point analysis, and toxicity assessment. Examples of these applications are listed in Table 6. The range of tested contaminants included chlorinated compounds, aromatic hydrocarbons, and nonpolar toxicants. The combination of microbiological and ecological knowledge, biochemical mechanisms, and field engineering designs are essential elements for successful in situ bioremediation using genetically modified bacteria.

Table 6: Application of genetically modified bacteria for assessing the biodegradation process efficiency

Microorganisms	Application	Contaminants	Reference
A. eutrophus H850Lr	process monitoring	PCB	Van Dyke et al. [107]
P. putida TVA8	process monitoring	TCE, BTEX	Applegate et al. [108]
P. fluorescens HK44	process monitoring	naphthalene, anthracene	Sayler and Ripp [109]
B. cepacia BRI6001L	strain monitoring	2,4-D	Masson et al. [110]
P. fluorescens 10586s/pUCD607	stress response	BTEX	Sousa et al. [111]
Pseudomonas strain Shk1	toxicity assessment	2, 4-dinitrophenol hydroquinone	Kelly et al. [112]

| A. eutrophus 2050 | end point analysis | non polar narcotics | Layton et al. [113] |

CONCLUSIONS

Cleaning up of petroleum hydrocarbons in the subsurface environment is a real world problem. A better understanding of the mechanism of biodegradation has a high ecological significance that depends on the indigenous microorganisms to transform or mineralize the organic contaminants. Microbial degradation process aids the elimination of spilled oil from the environment after critical removal of large amounts of the oil by various physical and chemical methods. This is possible because microorganisms have enzyme systems to degrade and utilize different hydrocarbons as a source of carbon and energy.

The use of genetically modified (GM) bacteria represents a research frontier with broad implications. The potential benefits of using genetically modified bacteria are significant. But the need for GM bacteria may be questionable for many cases, considering that indigenous species often perform adequately but we do not tap the full potential of wild species due to our limited understanding of various phytoremediation mechanisms, including the regulation of enzyme systems that degrade pollutants.

Therefore, based on the present review, it may be concluded that microbial degradation can be considered as a key component in the cleanup strategy for petroleum hydrocarbon remediation.

REFERENCES

1. K. A. Kvenvolden and C. K. Cooper, "Natural seepage of crude oil into the marine environment," Geo-Marine Letters, vol. 23, no. 3-4, pp. 140–146, 2003.
2. C. Holliger, S. Gaspard, G. Glod, C. Heijman, W. Schumacher, R. P. Schwarzenbach, and F. Vazquez, "Contaminated environments in the subsurface and bioremediation: organic contaminants," FEMS Microbiology Reviews, vol. 20, no. 3-4, pp. 517–523, 1997.

3. P. J. J. Alvarez and T. M. Vogel, "Substrate interactions of benzene, toluene, and para-xylene during microbial degradation by pure cultures and mixed culture aquifer slurries," Applied and Environmental Microbiology, vol. 57, no. 10, pp. 2981–2985, 1991.
4. J. I. Medina-Bellver, P. Marín, A. Delgado, A. Rodríguez-Sánchez, E. Reyes, J. L. Ramos, and S. Marqués, "Evidence for in situ crude oil biodegradation after the Prestige oil spill," Environmental Microbiology, vol. 7, no. 6, pp. 773–779, 2005.
5. T. M. April, J. M. Foght, and R. S. Currah, "Hydrocarbon-degrading filamentous fungi isolated from flare pit soils in northern and western Canada," Canadian Journal of Microbiology, vol. 46, no. 1, pp. 38–49, 2000.
6. W. Ulrici, "Contaminant soil areas, different countries and contaminant monitoring of contaminants," in Environmental Process II. Soil Decontamination Biotechnology, H. J. Rehm and G. Reed, Eds., vol. 11, pp. 5–42, 2000.
7. J. G. Leahy and R. R. Colwell, "Microbial degradation of hydrocarbons in the environment," Microbiological Reviews, vol. 54, no. 3, pp. 305–315, 1990.
8. C. E. Zobell, "Action of microorganisms on hydrocarbons," Bacteriological Reviews, vol. 10, pp. 1–49, 1946.
9. R. M. Atlas, "Microbial degradation of petroleum hydrocarbons: an environmental perspective," Microbiological Reviews, vol. 45, no. 1, pp. 180–209, 1981.
10. R. M. Atlas, Ed., Petroleum Microbiology, Macmillion, New York, NY, USA, 1984.
11. R. M. Atlas and R. Bartha, "Hydrocabon biodegradation and oil spill bioremediation," Advances in Microbial Ecology, vol. 12, pp. 287–338, 1992.
12. J. M. Foght and D. W. S. Westlake, "Biodegradation of hydrocarbons in freshwater," in Oil in Freshwater: Chemistry, Biology, Countermeasure Technology, J. H. Vandermeulen and S. R. Hrudey, Eds., pp. 217–230, Pergamon Press, New York, NY, USA, 1987.
13. R. M. Atlas and R. Bartha, "Fundamentals and applications," in Microbial Ecology, pp. 523–530, Benjamin/Cummings, San

Francisco, Calif, USA, 4th edition, 1998.

14. A. J. Mearns, "Cleaning oiled shores: putting bioremediation to the test," Spill Science and Technology Bulletin, vol. 4, no. 4, pp. 209–217, 1997.

15. R. C. Prince, "Petroleum spill bioremediation in marine environments," Critical Reviews in Microbiology, vol. 19, no. 4, pp. 217–242, 1993.

16. R. P. J. Swannell, K. Lee, and M. Mcdonagh, "Field evaluations of marine oil spill bioremediation,"Microbiological Reviews, vol. 60, no. 2, pp. 342–365, 1996.

17. A. D. Venosa, M. T. Suidan, and M. T. Suidan, "Bioremediation of an experimental oil spill on the shoreline of Delaware Bay," Environmental Science and Technology, vol. 30, no. 5, pp. 1764–1775, 1996.

18. A. D. Venosa, D. W. King, and G. A. Sorial, "The baffled flask test for dispersant effectiveness: a round Robin evaluation of reproducibility and repeatability," Spill Science and Technology Bulletin, vol. 7, no. 5-6, pp. 299–308, 2002.

19. R. R. Colwell, J. D. Walker, and J. J. Cooney, "Ecological aspects of microbial degradation of petroleum in the marine environment," Critical Reviews in Microbiology, vol. 5, no. 4, pp. 423–445, 1977.

20. J. J. Cooney, S. A. Silver, and E. A. Beck, "Factors influencing hydrocarbon degradation in three freshwater lakes," Microbial Ecology, vol. 11, no. 2, pp. 127–137, 1985.

21. S. Barathi and N. Vasudevan, "Utilization of petroleum hydrocarbons by Pseudomonas fluorescensisolated from a petroleum-contaminated soil," Environment International, vol. 26, no. 5-6, pp. 413–416, 2001.

22. J. J. Perry, "Microbial metabolism of cyclic alkanes," in Petroleum Microbiology, R. M. Atlas, Ed., pp. 61–98, Macmillan, New York, NY, USA, 1984.

23. R. Atlas and J. Bragg, "Bioremediation of marine oil spills: when and when not—the Exxon Valdez experience," Microbial Biotechnology, vol. 2, no. 2, pp. 213–221, 2009.

24. R. M. Atlas, "Petroleum microbiology," in Encyclopedia of Microbiology, pp. 363–369, Academic Press, Baltimore, Md,

USA, 1992.
25. O. O. Amund and N. Nwokoye, "Hydrocarbon potentials of yeast isolates from a polluted Lagoon," Journal of Scientific Research and Development, vol. 1, pp. 65–68, 1993.
26. B. Lal and S. Khanna, "Degradation of crude oil by Acinetobacter calcoaceticus and Alcaligenes odorans," Journal of Applied Bacteriology, vol. 81, no. 4, pp. 355–362, 1996.
27. D. M. Jones, A. G. Douglas, R. J. Parkes, J. Taylor, W. Giger, and C. Schaffner, "The recognition of biodegraded petroleum-derived aromatic hydrocarbons in recent marine sediments," Marine Pollution Bulletin, vol. 14, no. 3, pp. 103–108, 1983.
28. S. A. Adebusoye, M. O. Ilori, O. O. Amund, O. D. Teniola, and S. O. Olatope, "Microbial degradation of petroleum hydrocarbons in a polluted tropical stream," World Journal of Microbiology and Biotechnology, vol. 23, no. 8, pp. 1149–1159, 2007.
29. J. Jones, M. Knight, and J. A. Byron, "Effect of gross population by kerosene hydrocarbons on the microflora of a moorland soil," Nature, vol. 227, p. 1166, 1970.
30. Y. Pinholt, S. Struwe, and A. Kjoller, "Microbial changes during oil decomposition in soil," Holarctic Ecology, vol. 2, pp. 195–200, 1979.
31. S. L. Hollaway, G. M. Faw, and R. K. Sizemore, "The bacterial community composition of an active oil field in the Northwestern Gulf of Mexico," Marine Pollution Bulletin, vol. 11, no. 6, pp. 153–156, 1980.
32. G. J. Mulkins Phillips and J. E. Stewart, "Distribution of hydrocarbon utilizing bacteria in Northwestern Atlantic waters and coastal sediments," Canadian Journal of Microbiology, vol. 20, no. 7, pp. 955–962, 1974.
33. R. Bartha and I. Bossert, "The treatment and disposal of petroleum wastes," in Petroleum Microbiology, R. M. Atlas, Ed., pp. 553–578, Macmillan, New York, NY, USA, 1984.
34. J. J. Cooney, "The fate of petroleum pollutants in fresh water ecosystems," in Petroleum Microbiology, R. M. Atlas, Ed., pp. 399–434, Macmillan, New York, NY, USA, 1984.
35. R. M. Atlas, "Effects of hydrocarbons on micro-organisms and biodegradation in Arctic ecosystems," in Petroleum Effects in the Arctic Environment, F. R. Engelhardt, Ed., pp. 63–99, Elsevier,

London, UK, 1985.

36. G. Floodgate, "The fate of petroleum in marine ecosystems," in Petroleum Microbiology, R. M. Atlas, Ed., pp. 355–398, Macmillion, New York, NY, USA, 1984.

37. K. S. M. Rahman, T. J. Rahman, Y. Kourkoutas, I. Petsas, R. Marchant, and I. M. Banat, "Enhanced bioremediation of n-alkane in petroleum sludge using bacterial consortium amended with rhamnolipid and micronutrients," Bioresource Technology, vol. 90, no. 2, pp. 159–168, 2003.

38. R. J. W. Brooijmans, M. I. Pastink, and R. J. Siezen, "Hydrocarbon-degrading bacteria: the oil-spill clean-up crew," Microbial Biotechnology, vol. 2, no. 6, pp. 587–594, 2009.

39. M. M. Yakimov, K. N. Timmis, and P. N. Golyshin, "Obligate oil-degrading marine bacteria," Current Opinion in Biotechnology, vol. 18, no. 3, pp. 257–266, 2007.

40. K. Das and A. K. Mukherjee, "Crude petroleum-oil biodegradation efficiency of Bacillus subtilis andPseudomonas aeruginosa strains isolated from a petroleum-oil contaminated soil from North-East India," Bioresource Technology, vol. 98, no. 7, pp. 1339–1345, 2007.

41. M. Throne-Holst, A. Wentzel, T. E. Ellingsen, H.-K. Kotlar, and S. B. Zotchev, "Identification of novel genes involved in long-chain n-alkane degradation by Acinetobacter sp. strain DSM 17874," Applied and Environmental Microbiology, vol. 73, no. 10, pp. 3327–3332, 2007.

42. F. Chaillan, A. Le Flèche, E. Bury, Y.-H. Phantavong, P. Grimont, A. Saliot, and J. Oudot, "Identification and biodegradation potential of tropical aerobic hydrocarbon-degrading microorganisms," Research in Microbiology, vol. 155, no. 7, pp. 587–595, 2004.

43. A. J. Daugulis and C. M. McCracken, "Microbial degradation of high and low molecular weight polyaromatic hydrocarbons in a two-phase partitioning bioreactor by two strains of Sphingomonas sp,"Biotechnology Letters, vol. 25, no. 17, pp. 1441–1444, 2003.

44. H. Singh, Mycoremediation: Fungal Bioremediation, Wiley-Interscience, New York, NY, USA, 2006.

45. E. Bogusławska-Was and W. D browski, "The seasonal variability of yeasts and yeast-like organisms in water and bottom sediment

of the Szczecin Lagoon," International Journal of Hygiene and Environmental Health, vol. 203, no. 5-6, pp. 451–458, 2001.
46. I. R. McDonald, C. B. Miguez, G. Rogge, D. Bourque, K. D. Wendlandt, D. Groleau, and J. Murrell, "Diversity of soluble methane monooxygenase-containing methanotrophs isolated from polluted environments," FEMS Microbiology Letters, vol. 255, no. 2, pp. 225–232, 2006.
47. B. Jan, V. Beilen, M. Neuenschwunder, T. H. M. Suits, C. Roth, S. B. Balada, and B. Witholt, "Rubredoxins involved in alkane degradation," The Journal of Bacteriology, vol. 184, no. 6, pp. 1722–1732, 2003.
48. T. Iida, T. Sumita, A. Ohta, and M. Takagi, "The cytochrome P450ALK multigene family of an n-alkane-assimilating yeast, Yarrowia lipolytica: cloning and characterization of genes coding for new CYP52 family members," Yeast, vol. 16, no. 12, pp. 1077–1087, 2000.
49. J. B. Van Beilen, E. G. Funhoff, and E. G. Funhoff, "Cytochrome P450 alkane hydroxylases of the CYP153 family are common in alkane-degrading eubacteria lacking integral membrane alkane hydroxylases," Applied and Environmental Microbiology, vol. 72, no. 1, pp. 59–65, 2006.
50. J. H. O. Maeng, Y. Sakai, Y. Tani, and N. Kato, "Isolation and characterization of a novel oxygenase that catalyzes the first step of n-alkane oxidation in Acinetobacter sp. strain M-1," Journal of Bacteriology, vol. 178, no. 13, pp. 3695–3700, 1996.
51. J. D. Walker, R. R. Colwell, Z. Vaituzis, and S. A. Meyer, "Petroleum degrading achlorophyllous algaPrototheca zopfii," Nature, vol. 254, no. 5499, pp. 423–424, 1975.
52. C. E. Cerniglia, D. T. Gibson, and C. Van Baalen, "Oxidation of naphthalene by cyanobacteria and microalgae," Journal of General Microbiology, vol. 116, no. 2, pp. 495–500, 1980.
53. M. L. Brusseau, "The impact of physical, chemical and biological factors on biodegradation," inProceedings of the International Conference on Biotechnology for Soil Remediation: Scientific Bases and Practical Applications, R. Serra, Ed., pp. 81–98, C.I.P.A. S.R.L., Milan, Italy, 1998.
54. R. M. Atlas, "Effects of temperature and crude oil composition

on petroleum biodegradation," Journal of Applied Microbiology, vol. 30, no. 3, pp. 396–403, 1975.

55. A. Daverey and K. Pakshirajan, "Production of sophorolipids by the yeast Candida bombicola using simple and low cost fermentative media," Food Research International, vol. 42, no. 4, pp. 499–504, 2009.

56. M. Kumar, V. León, A. De Sisto Materano, O. A. Ilzins, and L. Luis, "Biosurfactant production and hydrocarbon-degradation by halotolerant and thermotolerant Pseudomonas sp," World Journal of Microbiology and Biotechnology, vol. 24, no. 7, pp. 1047–1057, 2008.

57. K. Muthusamy, S. Gopalakrishnan, T. K. Ravi, and P. Sivachidambaram, "Biosurfactants: properties, commercial production and application," Current Science, vol. 94, no. 6, pp. 736–747, 2008.

58. A. Mahmound, Y. Aziza, A. Abdeltif, and M. Rachida, "Biosurfactant production by Bacillus strain injected in the petroleum reservoirs," Journal of Industrial Microbiology & Biotechnology, vol. 35, pp. 1303–1306, 2008.

59. N. Youssef, D. R. Simpson, K. E. Duncan, M. J. McInerney, M. Folmsbee, T. Fincher, and R. M. Knapp, "In situ biosurfactant production by Bacillus strains injected into a limestone petroleum reservoir,"Applied and Environmental Microbiology, vol. 73, no. 4, pp. 1239–1247, 2007.

60. M. O. Ilori, C. J. Amobi, and A. C. Odocha, "Factors affecting biosurfactant production by oil degradingAeromonas spp. isolated from a tropical environment," Chemosphere, vol. 61, no. 7, pp. 985–992, 2005.

61. A. Tabatabaee, M. M. Assadi, A. A. Noohi, and V. A. Sajadian, "Isolation of biosurfactant producing bacteria from oil reservoirs," Iranian Journal of Environmental Health Science & Engineering, vol. 2, no. 1, pp. 6–12, 2005.

62. J. M. Foght, D. W. S. Westlake, W. M. Johnson, and H. F. Ridgway, "Environmental gasoline-utilizing isolates and clinical isolates of Pseudomonas aeruginosa are taxonomically indistinguishable by chemotaxonomic and molecular techniques," Microbiology, vol. 142, no. 9, pp. 2333–2340, 1996.

63. A. D. Venosa and X. Zhu, "Biodegradation of crude oil contaminating marine shorelines and freshwater wetlands," Spill Science and Technology Bulletin, vol. 8, no. 2, pp. 163–178, 2003.

64. E. Pelletier, D. Delille, and B. Delille, "Crude oil bioremediation in sub-Antarctic intertidal sediments: chemistry and toxicity of oiled residues," Marine Environmental Research, vol. 57, no. 4, pp. 311–327, 2004.

65. D. Delille, F. Coulon, and E. Pelletier, "Effects of temperature warming during a bioremediation study of natural and nutrient-amended hydrocarbon-contaminated sub-Antarctic soils," Cold Regions Science and Technology, vol. 40, no. 1-2, pp. 61–70, 2004.

66. W. J. Mitsch and J. G. Gosselink, Wetlands, John Wiley & Sons, New York, NY, USA, 2nd edition, 1993.

67. S.-C. Choi, K. K. Kwon, J. H. Sohn, and S.-J. Kim, "Evaluation of fertilizer additions to stimulate oil biodegradation in sand seashore mesocosms," Journal of Microbiology and Biotechnology, vol. 12, no. 3, pp. 431–436, 2002.

68. S.-J. Kim, D. H. Choi, D. S. Sim, and Y.-S. Oh, "Evaluation of bioremediation effectiveness on crude oil-contaminated sand," Chemosphere, vol. 59, no. 6, pp. 845–852, 2005.

69. F. Chaillan, C. H. Chaîneau, V. Point, A. Saliot, and J. Oudot, "Factors inhibiting bioremediation of soil contaminated with weathered oils and drill cuttings," Environmental Pollution, vol. 144, no. 1, pp. 255–265, 2006.

70. J. Oudot, F. X. Merlin, and P. Pinvidic, "Weathering rates of oil components in a bioremediation experiment in estuarine sediments," Marine Environmental Research, vol. 45, no. 2, pp. 113–125, 1998.

71. C. H. Chaîneau, G. Rougeux, C. Yéprémian, and J. Oudot, "Effects of nutrient concentration on the biodegradation of crude oil and associated microbial populations in the soil," Soil Biology and Biochemistry, vol. 37, no. 8, pp. 1490–1497, 2005.

72. L. M. Carmichael and F. K. Pfaender, "The effect of inorganic and organic supplements on the microbial degradation of phenanthrene and pyrene in soils," Biodegradation, vol. 8, no. 1, pp. 1–13, 1997.

73. J. C. Okolo, E. N. Amadi, and C. T. I. Odu, "Effects of soil treatments containing poultry manure on crude oil degradation in a sandy loam soil," Applied Ecology and Environmental Research, vol. 3, no. 1, pp. 47–53, 2005.

74. H. Maki, T. Sasaki, and S. Haramaya, "Photooxidation of biodegradable crude oil and toxicity of the photooxidized products," Chemosphere, vol. 44, pp. 1145–1151, 2005.

75. W. Fritsche and M. Hofrichter, "Aerobic degradation by microorganisms," in Environmental Processes- Soil Decontamination, J. Klein, Ed., pp. 146–155, Wiley-VCH, Weinheim, Germany, 2000.

76. R. K. Hommel, "Formation and phylogenetic role of biosurfactants," Journal of Applied Microbiology, vol. 89, no. 1, pp. 158–119, 1990.

77. J. B. Van Beilen and E. G. Funhoff, "Alkane hydroxylases involved in microbial alkane degradation," Applied Microbiology and Biotechnology, vol. 74, no. 1, pp. 13–21, 2007.

78. T. Zimmer, M. Ohkuma, A. Ohta, M. Takagi, and W.-H. Schunck, "The CYP52 multigene family ofCandida maltosa encodes functionally diverse n-alkane-inducible cytochromes p450," Biochemical and Biophysical Research Communications, vol. 224, no. 3, pp. 784–789, 1996.

79. U. Scheuer, T. Zimmer, D. Becher, F. Schauer, and W.-H. Schunck, "Oxygenation cascade in conversion of n-alkanes to , -dioic acids catalyzed by cytochrome P450 52A3," Journal of Biological Chemistry, vol. 273, no. 49, pp. 32528–32534, 1998.

80. J. B. Van Beilen and E. G. Funhoff, "Expanding the alkane oxygenase toolbox: new enzymes and applications," Current Opinion in Biotechnology, vol. 16, no. 3, pp. 308–314, 2005.

81. M. O. Ilori, S. A. Adebusoye, and A. C. Ojo, "Isolation and characterization of hydrocarbon-degrading and biosurfactant-producing yeast strains obtained from a polluted lagoon water," World Journal of Microbiology and Biotechnology, vol. 24, no. 11, pp. 2539–2545, 2008.

82. G. S. Kiran, T. A. Hema, R. Gandhimathi, J. Selvin, T. A. Thomas, T. Rajeetha Ravji, and K. Natarajaseenivasan, "Optimization and production of a biosurfactant from the sponge-associated marine

fungus Aspergillus ustus MSF3," Colloids and Surfaces B, vol. 73, no. 2, pp. 250–256, 2009.

83. O. S. Obayori, M. O. Ilori, S. A. Adebusoye, G. O. Oyetibo, A. E. Omotayo, and O. O. Amund, "Degradation of hydrocarbons and biosurfactant production by Pseudomonas sp. strain LP1," World Journal of Microbiology and Biotechnology, vol. 25, no. 9, pp. 1615–1623, 2009.

84. M. L. Brusseau, R. M. Miller, Y. Zhang, X. Wang, and G. Y. Bai, "Biosurfactant and cosolvent enhanced remediation of contaminated media," ACS Symposium Series, vol. 594, pp. 82–94, 1995.

85. G. Bai, M. L. Brusseau, and R. M. Miller, "Biosurfactant-enhanced removal of residual hydrocarbon from soil," Journal of Contaminant Hydrology, vol. 25, no. 1-2, pp. 157–170, 1997.

86. T. Barkay, S. Navon-Venezia, E. Z. Ron, and E. Rosenberg, "Enhancement of solubilization and biodegradation of polyaromatic hydrocarbons by the bioemulsifier alasan," Applied and Environmental Microbiology, vol. 65, no. 6, pp. 2697–2702, 1999.

87. S. S. Cameotra and P. Singh, "Bioremediation of oil sludge using crude biosurfactants," International Biodeterioration and Biodegradation, vol. 62, no. 3, pp. 274–280, 2008.

88. R. Beal and W. B. Betts, "Role of rhamnolipid biosurfactants in the uptake and mineralization of hexadecane in Pseudomonas aeruginosa," Journal of Applied Microbiology, vol. 89, no. 1, pp. 158–168, 2000.

89. O. Pornsunthorntawee, P. Wongpanit, S. Chavadej, M. Abe, and R. Rujiravanit, "Structural and physicochemical characterization of crude biosurfactant produced by Pseudomonas aeruginosa SP4 isolated from petroleum—contaminated soil," Bioresource Technology, vol. 99, no. 6, pp. 1589–1595, 2008.

90. M. Nikolopoulou and N. Kalogerakis, "Biostimulation strategies for fresh and chronically polluted marine environments with petroleum hydrocarbons," Journal of Chemical Technology and Biotechnology, vol. 84, no. 6, pp. 802–807, 2009.

91. N. G. Wilson and G. Bradley, "The effect of immobilization on rhamnolipid production by Pseudomonas fluorescens," Journal of

Applied Bacteriology, vol. 81, no. 5, pp. 525–530, 1996.

92. M. P. Díaz, K. G. Boyd, S. J. W. Grigson, and J. G. Burgess, "Biodegradation of crude oil across a wide range of salinities by an extremely halotolerant bacterial consortium MPD-M, immobilized onto polypropylene fibers," Biotechnology and Bioengineering, vol. 79, no. 2, pp. 145–153, 2002.

93. C. J. Cunningham, I. B. Ivshina, V. I. Lozinsky, M. S. Kuyukina, and J. C. Philp, "Bioremediation of diesel-contaminated soil by microorganisms immobilised in polyvinyl alcohol," International Biodeterioration and Biodegradation, vol. 54, no. 2-3, pp. 167–174, 2004.

94. R. N. Z. A. Rahman, F. M. Ghazali, A. B. Salleh, and M. Basri, "Biodegradation of hydrocarbon contamination by immobilized bacterial cells," Journal of Microbiology, vol. 44, no. 3, pp. 354–359, 2006.

95. W. J. Nichols, "The U.S. Environmental Protect Agency: National Oil and Hazardous Substances Pollution Contingency Plan, Subpart J Product Schedule (40 CFR 300.900)," in Proceedings of the International Oil Spill Conference, pp. 1479–1483, American Petroleum Institute, Washington, DC, USA, 2001.

96. R. Z. Hoff, "Bioremediation: an overview of its development and use for oil spill cleanup," Marine Pollution Bulletin, vol. 26, no. 9, pp. 476–481, 1993.

97. U.S. EPA, "Spill NCP Product Schedule," 2002, http://www.epa.gov/oilspill.

98. K. Lee, G. H. Tremblay, J. Gauthier, S. E. Cobanli, and M. Griffin, "Bioaugmentation and biostimulation: a paradox between laboratory and field results," in Proceedings of the International Oil Spill Conference, pp. 697–705, American Petroleum Institute, Washington, DC, USA, 1997.

99. A. Ladousse and B. Tramier, "Results of 12 years of research in spilled oil bioremediation: inipol EAP 22," in Proceedings of the International Oil Spill Conference, pp. 577–581, American Petroleum Institute, Washington, DC, USA, 1991.

100. J. L. Ramos, A. Wasserfallen, K. Rose, and K. N. Timmis, "Redesigning metabolic routes: manipulation of TOL plasmid pathway for catabolism of alkylbenzoates," Science, vol. 235, no. 4788, pp. 593–596, 1987.

101. S. Panke, J. M. Sánchez-Romero, and V. De Lorenzo, "Engineering of quasi-natural Pseudomonas putidastrains toluene metabolism through an ortho-cleavage degradation pathway," Applied and Environmental Microbiology, vol. 64, no. 2, pp. 748–751, 1998.
102. F. Rojo, D. H. Pieper, K.-H. Engesser, H.-J. Knackmuss, and K. N. Timmis, "Assemblage of ortho cleavage route for simultaneous degradation of chloro- and methylaromatics," Science, vol. 238, no. 4832, pp. 1395–1398, 1987.
103. Y. Hrywna, T. V. Tsoi, O. V. Maltseva, J. F. Quensen III, and J. M. Tiedje, "Construction and characterization of two recombinant bacteria that grow on ortho- and para-substituted chlorobiphenyls," Applied and Environmental Microbiology, vol. 65, no. 5, pp. 2163–2169, 1999.
104. B. D. Erickson and F. J. Mondello, "Enhanced biodegradation of polychlorinated biphenyls after site-directed mutagenesis of a biphenyl dioxygenase gene," Applied and Environmental Microbiology, vol. 59, no. 11, pp. 3858–3862, 1993.
105. A. Suyama, R. Iwakiri, N. Kimura, A. Nishi, K. Nakamura, and K. Furukawa, "Engineering hybridpseudomonads capable of utilizing a wide range of aromatic hydrocarbons and of efficient degradation of trichloroethylene," Journal of Bacteriology, vol. 178, no. 14, pp. 4039–4046, 1996.
106. T. C. Zwick, E. A. Foote, A. J. Pollack, et al., "Effects of nutrient addition during bioventing of fuel contaminated soils in an arid environment," in In-Situ and On-Site Bioremediation, pp. 403–409, Battelle Press, Columbus, Ohio, USA, 1997.
107. M. I. Van Dyke, H. Lee, and J. T. Trevors, "Survival of luxAB-marked Alcaligenes eutrophus H850 in PCB-contaminated soil and sediment," Journal of Chemical Technology and Biotechnology, vol. 65, no. 2, pp. 115–122, 1996.
108. B. M. Applegate, S. R. Kehrmeyer, and G. S. Sayler, "A chromosomally based tod-luxCDABE whole-cell reporter for benzene, toluene, ethybenzene, and xylene (BTEX) sensing," Applied and Environmental Microbiology, vol. 64, no. 7, pp. 2730–2735, 1998.
109. G. S. Sayler and S. Ripp, "Field applications of genetically engineered microorganisms for bioremediation processes," Current Opinion in Biotechnology, vol. 11, no. 3, pp. 286–289,

2000.

110. L. Masson, B. E. Tabashnik, A. Mazza, G. Préfontaine, L. Potvin, R. Brousseau, and J.-L. Schwartz, "Mutagenic analysis of a conserved region of domain III in the Cry1ac toxin of Bacillus thuringiensis," Applied and Environmental Microbiology, vol. 68, no. 1, pp. 194–200, 2002.

111. C. Sousa, V. De Lorenzo, and A. Cebolla, "Modulation of gene expression through chromosomal positioning in Escherichia coli," Microbiology, vol. 143, no. 6, pp. 2071–2078, 1997.

112. C. J. Kelly, C. A. Lajoie, A. C. Layton, and G. S. Sayler, "Bioluminescent reporter bacterium for toxicity monitoring in biological wastewater treatment systems," Water Environment Research, vol. 71, no. 1, pp. 31–35, 1999.

113. A. C. Layton, B. Gregory, T. W. Schultz, and G. S. Sayler, "Validation of genetically engineered bioluminescent surfactant resistant bacteria as toxicity assessment tools," Ecotoxicology and Environmental Safety, vol. 43, no. 2, pp. 222–228, 1999.

114. S. Le Floch, F.-X. Merlin, M. Guillerme, C. Dalmazzone, and P. Le Corre, "A field experimentation on bioremediation: bioren," Environmental Technology, vol. 20, no. 8, pp. 897–907, 1999.

115. S. Le Floch, F. X. Merlin, M. Guillerme, et al., "Bioren: recent experiment on oil polluted shoreline in temperate climate," in In-Situ and On-Site Bioremediation, pp. 411–417, Battelle Press, Columbus, Ohio, USA, 1997.

116. J. L. Schnoor, L. A. Licht, S. C. McCutcheon, N. L. Wolfe, and L. H. Carreira, "Phytoremediation of organic and nutrient contaminants," Environmental Science and Technology, vol. 29, no. 7, pp. 318A–323A, 1995.

117. D. Hecht and G. Badiane, "Phytoremediation," New Internationalist, June 1998.

118. K. V. Nedunuri, R. S. Govindaraju, M. K. Banks, A. P. Schwab, and Z. Chen, "Evaluation of phytoremediation for field-scale degradation of total petroleum hydrocarbons," Journal of Environmental Engineering, vol. 126, no. 6, pp. 483–490, 2000.

119. U.S. Army Corps of Engineers, Agriculturally Based Bioremediation of Petroleum-Contaminated Soils and Shallow Groundwater in Pacific Island Ecosystems, 2003.

120. R. K. Miya and M. K. Firestone, "Enhanced phenanthrene biodegradation in soil by slender oat root exudates and root debris," Journal of Environmental Quality, vol. 30, no. 6, pp. 1911–1918, 2001.

Chapter 8

Pore Structure and Limit Pressure of Gas Slippage Effect in Tight Sandstone

Lijun You, Kunlin Xue, Yili Kang, Yi Liao, and Lie Kong

State Key Laboratory of Oil and Gas Reservoir Geology and Exploitation, Southwest Petroleum University, Chengdu 610500, China

ABSTRACT

Gas slip effect is an important mechanism that the gas flow is different from liquid flow in porous media. It is generally considered that the lower the permeability in porous media is, the more severe slip effect of gas flow will be. We design and then carry out experiments with the increase of backpressure at the outlet of the core samples based on the definition of gas slip effect and in view of different levels of permeability of tight sandstone reservoir. This study inspects a limit pressure of the gas slip effect in tight sandstones and analyzes the characteristic parameter of capillary pressure curves. The experimental results indicate that gas slip effect can be eliminated when the backpressure reaches a limit pressure. When the backpressure exceeds the limit pressure, the measured gas permeability is a relatively stable value whose range is less than 3% for a given core sample. It is also found that the limit pressure increases with the decreasing in permeability and has close relation with pore structure of the core samples. The results have an important influence on correlation study on gas flow in porous medium, and are beneficial to reduce the workload of laboratory experiment.

INTRODUCTION

With the development of oil and gas exploration technology, tight gas reservoirs, the most realistic unconventional reservoirs, play and will continually play an increasingly vital role in gas reserves and supply [1]. According to the third resource assessment, tight sandstone gas resources in China are about $20 \times 10^{12}\,m^3$. Tight sandstone reservoirs face the huge difficulty of the exploitation because of slim throat, low porosity, low permeability, high content of clay mineral, and high capillary pressure. Gas slip effect, a phenomenon that will occur when gas flowing through a thin capillary tube or a fine porous medium, controls gas flow behavior and severely affects the ability of gas flow in tight sandstone gas reservoir. During this process, the velocity of gas in velocity layer in the immediate vicinity of the solid walls of the capillary or porous medium is not zero, which will cause an increase in gas flow rate in porous media [2–5].

Klinkenberg (1941) was the first to introduce the concept of gas slip effect into gas permeability measurement; the mathematical expression was given as [5]

$$K_a = K_\infty \left(1 + \frac{b}{p_m}\right).$$

(1)

K_a is gas permeability, μm^2.
K_∞ is Klinkenberg permeability, μm^2.
P_m is mean pressure, MPa.
b is gas slip factor, affected by pressure, temperature, pore structure of porous medium, and type of gas. The expression was given as

$$b = \frac{4C\lambda p_m}{r} \qquad (2)$$

λ is mean free path of gas molecules, mm.
r is radius of a capillary or a pore, mm.
C is constant.

It is indicated by (2) that gas slip factor is inversely proportional to radius of capillary.

According to Darcy's law, the expression of gas permeability is given as

$$K_a = \frac{2Qp_0\mu L}{A(p_1^2 - p_2^2)} \times 10^{-1} \qquad (3)$$

Q is volumetric flow rate, cm³/s.
μ is dynamic viscosity of the fluid, mPa·s.
A is cross-sectional area, cm².
L is length of core sample, cm.

Transforming (3)

$$Q\mu = \frac{5K_a A}{p_0} \times \frac{p_1^2 - p_2^2}{L}. \qquad (4)$$

From (4), the relationship between $Q\mu$ and $(p_1^2 - p_2^2)$ is a straight line with increasing backpressure. $(p_1^2 - p_2^2)$ is composed of two items.

One means the pressure drop across core sample $(p_1 + p_2)$ and the other is double the pore pressures of core sample $(p_1 - p_2)$ a. When $Q\mu$ and $(p_1^2 - p_2^2)$ are linear relationships and the gradient is a stable value, gas permeability is equal to Klinkenberg permeability. And the gas slip effect would be reduced with the increasing of inlet pressure [6].

Both permeability and gas reservoir pressure determine the extent of slippage effect impacting volumetric flow rate [7–9]. The lower the permeability and gas pressure are, the more prominent gas slip effect would be [10]. The influence factors of gas slip effect include permeability, pore pressure, and water saturation. Gas slip effect would be prominent when the permeability is less than 0.1×10^{-3} μm² and the pore pressure is a low value, while the specific boundaries of water saturation are not clear [11]. Gas slip factor is related to the pore structure [12].

Slippage effect affects gas production. In laboratory, gas permeability is usually measured at a succession of pressures to obtain the Klinkenberg permeability by correcting because of Slippage effect. Laboratory working is increased. Equation (1) suggests that gas permeability is equal to Klinkenberg permeability when $b/p_m = 0$ Some researchers indicated that gas slip effect can be prevented by increasing pore pressure of high permeable core samples, but the study about tight sandstone is rare, and the results of tight sandstone are very different. Until Now there is no terminology to describe this phenomenon. In the past, some researchers exerted a big backpressure by rule of thumb to reduce Klinkenberg effect, which increases the pressure-bearing demand of experimental cardholder.

When gas permeability was close to Klinkenberg permeability by improving mean pressure to cause b/p_m to approach to zero, we define the pore pressure or backpressure at the outlet of the core sample as limit pressure.

If we know the limit pressure, we can measure permeability by exerting a backpressure which is equal to or a little greater than limit pressure to mitigate slippage effect on experimental results, such as the effect of gas velocity on gas permeability due to fine migration [13].

The impetus for this work was a concern that finding the relation between limit pressure of eliminating gas slippage effect and pore structure parameters can help obtain the limit pressure of specific pore structure rock.

EXPERIMENTAL SAMPLES AND PROCEDURES

Core Samples

In this study, the tight sandstone core samples, from Permian in Upper Paleozoic in Ordos basin, involve four permeability levels (<0.1 × 10^{-3} μm², (0.1~0.3) × 10^{-3} μm², (0.3~1) × 10^{-3} μm², and >1 × 10^{-3} μm²). Nitrogen is regarded as displacing medium. The schematic diagram of the experimental apparatus is shown in Figure 1. It mainly consists of a high pressure core holder, a high pressure nitrogen cylinder, a high pressure pump, a backpressure regulator (BPR), and a gas flowmeter.

Figure 1: Schematic diagram of the experimental apparatus.

Procedure

(1) Seven samples of four permeability levels (<0.1 × 10^{-3} μm², (0.1~0.3) × 10^{-3} μm², (0.3~1) × 10^{-3} μm², and >1 × 10^{-3} μm²) are selected in the experiments. SS-1 is an outcrop sample that is different from others. Before conducting the porosity and permeability test,

the core samples in this work are dried for more than 48 hours at 60°C. The basic parameters of samples are listed in Table 1. Figure 2 shows the relationship between porosity and permeability for core samples. The red ones are the samples in the experiments. (2) After a core sample is installed into the core holder, a confining pressure of 7 MPa is applied. Before flow tests, the core sample is needed to stay at this confining pressure for at least four hours to make sure that the stress equilibrium is reached. To start a test, the outlet pressure is set at a designed backpressure. In this test, the backpressure increases from 0 MPa and its differential ranges from 0.1 MPa to 0.2 MPa. (3) When the backpressure is fixed, the inlet pressure is increased by using the regulator of the nitrogen cylinder. The pressure difference between inlet and outlet is 0.5 MPa, 1.0 MPa, 1.5 MPa, 2.0 MPa, and 2.5 MPa, respectively. Once a steady flow is reached, the gas flow rate at different pressures is recorded and the permeability is calculated. (4) Increase backpressure and repeat step (3). (5) Analyze the experiment data and illustrate $Q\mu$ versus $(p_1^2 - p_2^2)/L$ plots and K_a versus $1/p_m$ plots.

Table 1: Basic parameters of core samples

Samples	L(mm)	D (mm)	Φ (%)	Ka (10−3 μm2)
SS-1	57.30	25.10	15.13	0.05
SS-2	77.66	25.02	6.689	0.074
SS-3	59.71	24.74	5.895	0.150
SS-4	63.95	24.73	1.106	0.226
SS-5	59.05	24.76	5.647	0.507
SS-6	62.85	24.73	10.099	1.090
SS-7	59.17	24.59	12.461	1.320

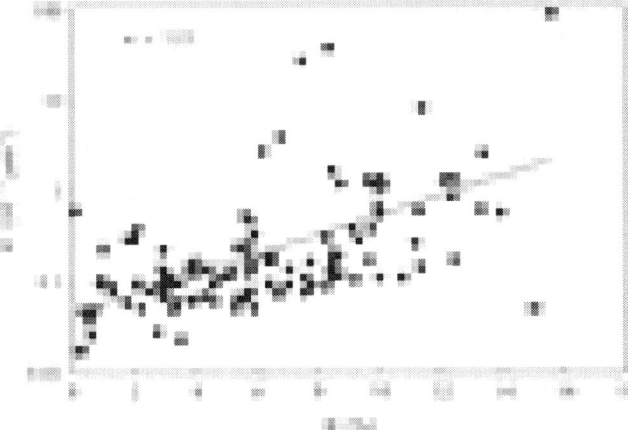

Figure 2: Relationships between porosity and permeability at 3 MPa for core samples.

RESULTS

Relationship between the Product of Flow Rate and Viscosity and Pressure Gradient

$Q\mu$ and $(p_1^2 - p_2^2)/L$ are plotted in Figures 3 and 4. It can be seen from Figures 3 and 4 that with the increasing of backpressure the slopes of the curves gradually reduce and the intercepts gradually approach to zero. When the backpressure reaches a specific level, the slope of the curve does not change with the pressure and intercept is equal to zero. The regression coefficient R^2 is more than 0.999. Equation (4) demonstrates that the gas slip effect can be eliminated when the permeability does not change with pressure. As shown in Figure 4, the relationship between $Q\mu$ and $(p_1^2 - p_2^2)/L$ is linear relation and the intercept is equal to zero when the backpressure at outlet reaches 0.9 MPa. The regression coefficient $R^2 = 0.9999$. When the outlet pressure of the sample exceeds 1 MPa, the curve is also fit for the law.

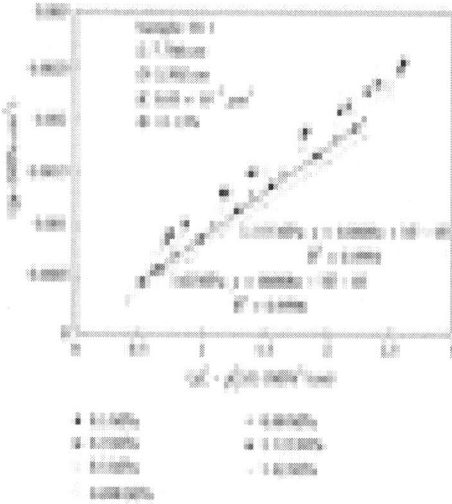

Figure 3: Relationships between $(p_1^2 - p_2^2)/L$ and $Q\mu$ at various backpressures for sample SS-1.

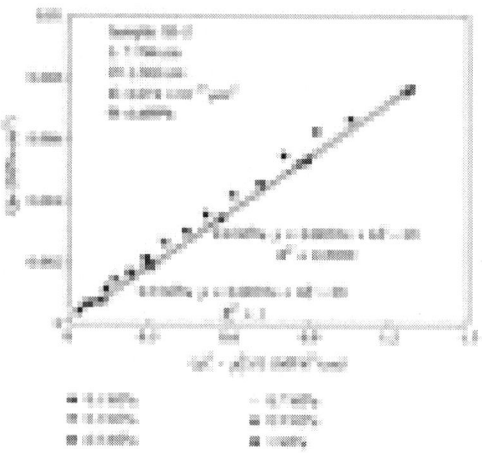

Figure 4: Relationships between $(p_1^2 - p_2^2)/L$ and $Q\mu$ at various backpressures for sample SS-2.

Relationship between Permeability and Reciprocal of Mean Pressure

Relationship between K_a and $1/p_m$ is presented in Figures 5 and 6. Figures 5 and 6 show that slip effect is obvious and permeability decreases with the increasing of mean pressure when the outlet pressure is atmospheric pressure. When the outlet pressure increases to a certain level, the relationship between K_a and $1/p_m$ is nearly horizontal and the gas permeability tested at different pressure drops is almost a stable value whose range is less than 3% and slip factor is less than 0.05 for a given sample (Table 2). The backpressure at outlet of the core sample is defined as limit pressure and the permeability is equal to liquid permeability.

Table 2: Basic parameters of core samples

Samples	(10–3 m2)	(%)	R35(m)	p_d (MPa)	limt (MPa)	Relationships between 1/ and for samples	b	range (%)
SS-1	0.05	15.130	0.023	1.8	1.42	= 0.0001 + 0.0032 = 0.0032 (1 + 0.031)	0.031	1.83
SS-2	0.074	6.689	0.226	2.38	0.9	= 0.0002 + 0.0231 = 0.0231 (1 + 0.008)	0.008	0.62
SS-3	0.150	5.895	0.382	2.15	0.6	= 0.0004 + 0.0382 = 0.0382 (1 + 0.010)	0.010	0.64
SS-4	0.226	1.106	2.061	1.47	0.5	= 0.0006 + 0.0771 = 0.0771 (1 + 0.008)	0.008	0.60
SS-5	0.507	5.647	0.811	0.912	0.45	= 0.0007 + 0.1534 = 0.1534 (1 + 0.005)	0.005	2.50
SS-6	1.090	10.099	0.769	0.85	0.4	= 0.0019 + 0.4162 = 0.4162 (1 + 0.005)	0.005	0.45
SS-7	1.320	12.461	0.718	0.62	0.35	= 0.0108 + 0.7911 = 0.7911 (1 + 0.013)	0.013	2.06

R_{35}: mean pore throat radius; p_d: displacement pressure; p_{limit}: limit pressure.

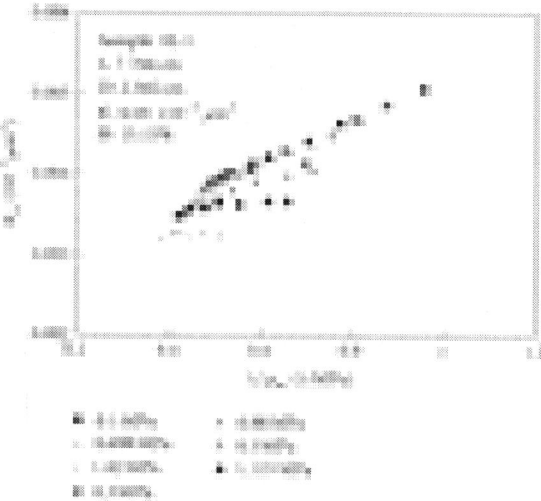

Figure 5: Relationships between $1/p_m$ and K_a at various backpressures for sample SS-1.

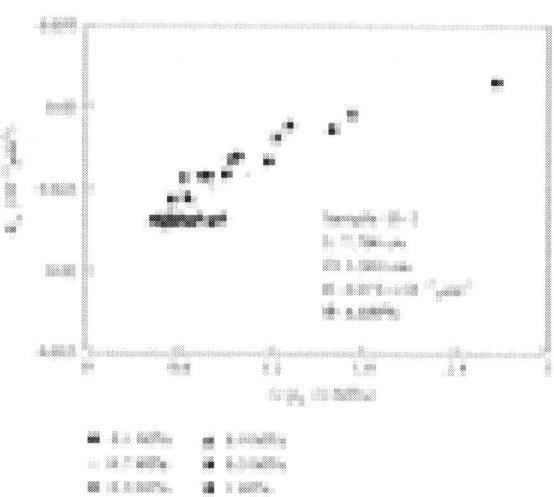

Figure 6: Relationships between $1/p_m$ and K_a at various backpressures for sample SS-2.

Gas slip factor *b* for sample SS-2 at different backpressure calculated from (1) is shown in Table 3. It can be seen from Table 3 that gas slip factor significantly reduced the increasing of backpressure and gas slip factor is less than 0.05 when the backpressure exceeds 0.9 MPa. At this case, gas slip effect can be eliminated.

Table 3: The influence of backpressure on gas slip factor for sample SS-2

Backpressure (MPa)	Relationships between n 1/pm and Ka for sample SS-2	b
0.1	y = 0.0072x + 0.0232 = 0.0232 (1 + 0.3103x)	0.3103
0.3	y = 0.0068x + 0.023 = 0.023 (1 + 0.2957x)	0.2957
0.5	y = 0.0061x + 0.0206 = 0.0206 (1 + 0.2961x)	0.2961
0.7	y = 0.0052x + 0.0205 = 0.0205 (1 + 0.2537x)	0.2537
0.9	y = 0.0002x + 0.0231 = 0.0231 (1 + 0.0086x)	0.0086
1	y = 0.0001x + 0.0227 = 0.0227 (1 + 0.0044x)	0.0044

DISCUSSION

- Gas Flow State in Tight Sandstone under Backpressure. Microstructure of tight sandstone is complicated, thereby Darcy's law only is not enough to describe the process of gas flow in micropore [14]. Gas flows in the different porous medium. Based on different mean free paths of gas molecules, the gas flow in micropore has different regions [15].

Knudsen (1934) introduced the concept of Knudsen number Kn, as is given by

$$Kn = \frac{\bar{\lambda}}{D}, \qquad (5)$$

where λ is mean free path of gas molecules and D is pore throat diameter

$$\bar{\lambda} = \frac{KT}{\sqrt{2}\pi d^2 P} \qquad (6)$$

Gas flow condition in micropore medium is decided by petrophysical property of the medium and mean free path of gas molecules [16, 17]. From the study of Liepmann, Stahl, and Kaviany et al. gas flow in tight sandstone is divided into three regions according to Knudsen number. It includes flow region, transition flow region, and viscous flow region.

Based on the results of Roy et al., gas flow in tight sandstone reservoir is divided by Knudsen number [18].

Ortega and Aguilera (2012) indicated that R_{35} in tight sandstone was the throat radius when the saturation was 35%. It can be defined as mean throat radius. Empirical formula is given as [19]

$$\log R_{35} = 0.732 + 0.588 \log K - 0.864 \log \phi \qquad (7)$$

Based on the porosity and permeability of core samples, R_{35} for SS-3 was calculated by (7), as shown in Table 2.

When the throat radius $R_{35}=0.382$ μm, the Knudsen number Kn at different outlet pressure was calculated by (5) and (6) as shown in Table 5.

From Table 4, Knudsen number Kn is greater than 0.001 and the gas slip effect is obvious when the pressure at outlet of core samples is atmospheric pressure. As the outlet pressure exceeds 0.6 MPa, Knudsen number Kn is less than 0.001 and the slip effect is negligible, which belongs to Darcy flow. Thus the gas slip effect can be neglected when the backpressure at outlet equals or exceeds the limit pressure.

Table 4: Knudsen number at different pressure for sample SS-3

2 = 0.1 MPa		2 = 0.6 MPa	
(MPa)	Kn	(MPa)	Kn
0.35	0.00376	0.85	0.00155
0.625	0.00211	1.1	0.00120
0.85	0.00155	1.35	0.00098
1.1	0.00120	1.6	0.00082
1.35	0.00098	1.85	0.00071

Table 5: Experimental results of limit pressure

Time	Author	(10−3 m2)	Gas	limit (MPa)
2009	Li et al. [20]	0.0053~0.25	N2	0.68~7.16
2007	Zhu et al. [24]	0.01~1	N2	0.5
2010	Gao et al. [23]	0.001~2	N2	1
2011	Ye et al. [21]	0.024~0.244	N2	<7

- Limit Pressure and Pore Structure. The experimental results of different permeability indicate that the limit pressure of tight sandstone decreases logarithmically with the increasing in permeability as well as in mean throat radius. The greater the permeability is, the smaller the range of limit pressure will reduce (Table 2, Figures 7 and 8).

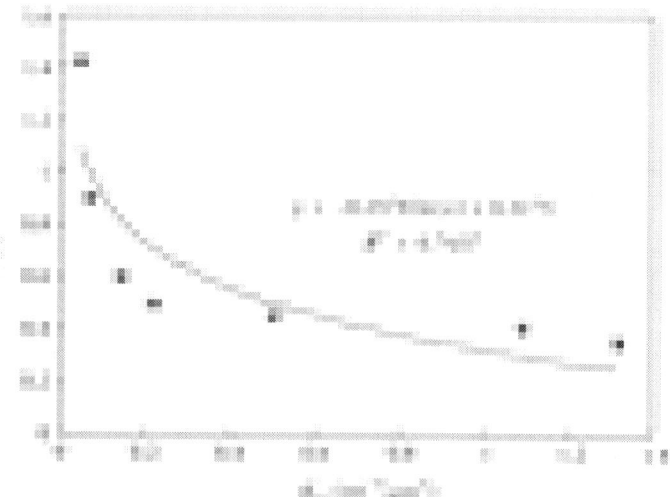

Figure 7: Relationships between K_a and p_{limi} for tight core samples.

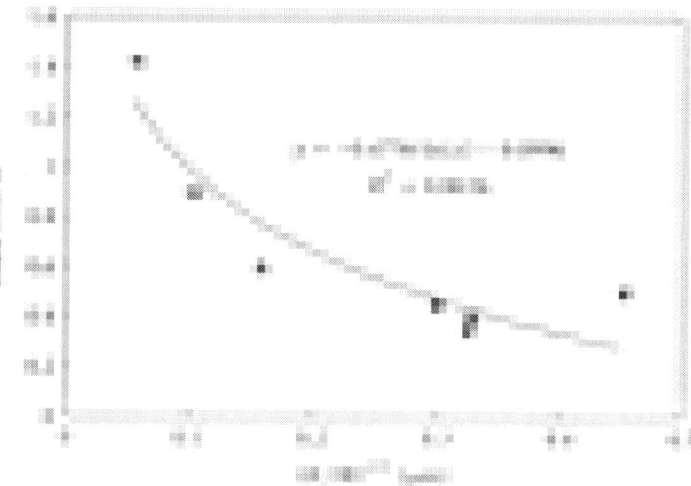

Figure 8: Relationships between integrated logistics index and p_{limi} for tight core samples.

For the experimented by Li et al. (2009), the limit pressure was confirmed as 0.68~7.16 MPa by increasing backpressure at outlet of core samples. The Empirical formula is given as [20]

$$p_{min} = -1.893 Ln K_\infty - 2.079 \tag{8}$$

The limit pressure from researchers has significant difference as shown in Table 5 [20–23]. The experimental results indicate that limit pressure is 0.35~1.5 MPa. It is close to the results of Zhu et al. (2007) [24] whose experiments also sampled from Permian in Upper Paleozoic in Ordos basin. The test results of this paper are validated by his result. It has been observed experimentally that pore structure has influence on gas slippage. In Figure 8, limit pressure and mean pore throat radius have logarithmic relation. The limit pressure reduced in logarithm with an increase in mean pore throat radius. From Figure 9, limit pressure of tight samples in Ordos basin is directly proportional to displacement pressure, and it is a quarter of displacement pressure. But the relations between limit pressure and displacement pressure are different from the other samples because of diverse pore structures. The limit pressure need quantitative study since it is an approximate value. The relation between pore structure parameters and limit pressure can be developed by fractal theory in porous medium [25, 26]. It is worth caring that, as limit pressure is associated with pore structure, the limit pressure of samples at different area needs to be tested by laboratory experiment.

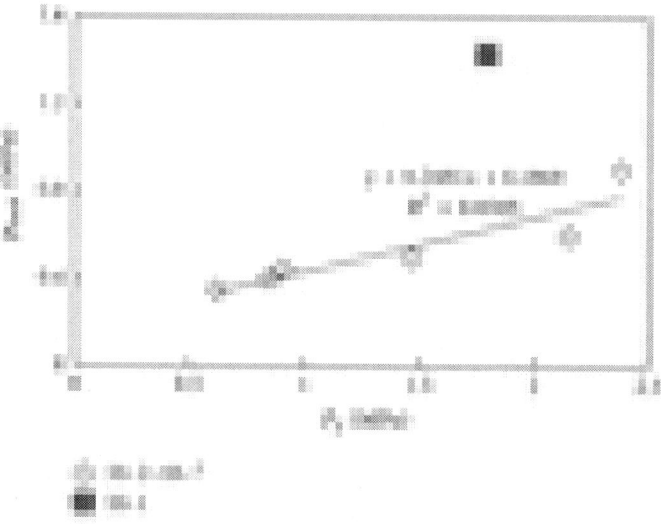

Figure 9: Relationships between displacement pressure and limit pressure for core samples.

CONCLUSIONS

- Limit Pressure. There exists gas slip effect in gas flow through tight sandstone, and exerting a certain backpressure can effectively reduce the gas slip effect. We define this backpressure as limit pressure.
- The Gas Slip Effect Is Negligible. When the backpressure equals or exceeds limit pressure, the gas permeability tested at different pressure drop is a stable value whose range is less than 3% and slip factor is less than 0.05 for a given sample. The gas slip effect is negligible and the permeability is equivalent to liquid permeability.
- There Are Close Relationship between the Limit Pressure and Pore Structure. The limit pressure of tight sandstone decreases logarithmically with the increasing of permeability and mean throat radius and is directly proportional to displacement pressure.

ACKNOWLEDGMENTS

This work was supported by the National Basic Research Program of China (2010CB226705), China Scholarship Fund, and Open Fund (PLN1117) of State Key Laboratory of Oil and Gas Reservoir Geology and Exploitation (Southwest Petroleum University).

REFERENCES

1. Y. Kang and P. Luo, "Current status and prospect of key techniques for exploration and production of tight sandstone gas reservoirs in China," Petroleum Exploration and Development, vol. 34, no. 2, pp. 239–245, 2007.
2. H. Krutter and R. J. Day, "Modification of permeability measurements," Oil Weekly, vol. 104, no. 4, pp. 24–32, 1941.
3. J. C. Calhoun and S. T. Yuster, "A study of the flow of homogeneous fluids through ideal porous media," in Proceedings of the Drilling and Production Practice, pp. 335–355, American Petroleum Institute, 1946.
4. J. Yang, Y. L. Kang, Q. G. Li, et al., "Characters of micro-structure and percolation in tight sandstone gas reservoirs," Advances in Mechanics, vol. 38, no. 2, pp. 229–235, 2008.
5. L. J. Klinkenberg, "The permeability of porous media to liquids and gases," in Proceedings of the Drilling and Production Practice, pp. 200–213, American Petroleum Institute, 1941.
6. Y. D. Yao, X. F. Li, J. L. Ge, and Z. Ning, "Experimental research for klinkenberg effect of gas percolation in low permeable gas reservoirs," Natural Gas Industry, vol. 24, no. 11, pp. 14–102, 2004.
7. V. Blanchard, D. Lasseux, H. Bertin et al., "Gas/water flow in porous media in the presence of adsorbed polymer: experimental study on non-darcy effects," in Proceedings of the 15th SPE-DOE Improved Oil Recovery Symposium: Old Reservoirs New Tricks A Global Perspective (SPE '06), pp. 529–538, April 2006.
8. W. Tanikawa and T. Shimamoto, "Klinkenberg effect for gas permeability and its comparison to water permeability for porous

sedimentary rocks," Hydrology and Earth System Sciences, vol. 3, pp. 1315–1338, 2006.
9. M. Tadayoni and M. Valadkhani, "New approach for the prediction of Klinkenberg permeability in situ for low permeability sandstone in tight gas reservoir," in Proceedings of the SPE Middle East Unconventional Gas Conference and Exhibition, 2012.
10. R. L. Luo, S. L. Cheng, H. Zhu, et al., "Problems on the study of slippage effect in low-permeability gas reservoirs," Gas Industry, vol. 27, no. 4, pp. 92–94, 2007.
11. J. Yan, N. S. Zhang, X. J. Liu, et al., "Research on the critical index of considering gas slippage effect," Journal of Wuhan Polytechnic University, vol. 28, no. 3, pp. 30–32, 2009.
12. A. Zeinijahromi, A. Vaz, and P. Bedrikovetsky, "Well impairment by fines migration in gas fields," Journal of Petroleum Science and Engineering, vol. 88-89, pp. 125–135, 2012.
13. W. F. Brace, J. B. Walsh, and W. T. Frangos, "Permeability of granite under high pressure," Journal of Geophysical Research, vol. 73, no. 6, pp. 2225–2236, 1968.
14. T. Ertekin, G. R. King, and F. C. Schwerer, "Dynamic gas slippage: a unique dual-mechanism approach to the flow of gas in tight formation," SPE Formation Evaluation, vol. 1, no. 1, pp. 43–52, 1986.
15. R. Rangarajan, M. A. Mazid, T. Matsuura, and S. Sourirajan, "Permeation of pure gases under pressure through asymmetric porous membranes. Membrane characterization and prediction of performance," Industrial & Engineering Chemistry Process Design and Development, vol. 23, no. 1, pp. 79–87, 1984.
16. F. Civan, "A triple-mechanism fractal model with hydraulic dispersion for gas permeation in tight reservoirs," in Proceedings of the SPE International Petroleum Conference and Exhibithion in Mexico, pp. 261–269, February 2002.
17. Y. S. Wu, K. Pruess, and P. Persoff, "Gas flow in porous media with Klinkenberg effects," Transport in Porous Media, vol. 32, no. 1, pp. 117–137, 1998.
18. S. Roy, R. Raju, H. F. Chuang, B. A. Cruden, and M. Meyyappan, "Modeling gas flow through microchannels and nanopores," Journal of Applied Physics, vol. 93, no. 8, pp. 4870–4879, 2003.

19. C. Ortega and R. Aguilera, "A complete petrophysical evaluation method for tight formations from only drill cuttings in the absence of well logs," in Proceedings of the SPE Canadian Unconventional Resources Conference, pp. 1–21, 2012.
20. S. Li, M. Dong, and Z. Li, "Measurement and revised interpretation of gas flow behavior in tight reservoir cores," Journal of Petroleum Science and Engineering, vol. 65, no. 1-2, pp. 81–88, 2009.
21. L. Y. Ye, S. S. Gao, W. Xiong, et al., "Percolation characteristics of gas in sandstone gas reservoir with low permeability under reservoir pressure," Complex Hydrocarbon Reservoirs, vol. 4, no. 1, pp. 59–62, 2011.
22. W. Xiong, S. S. Gao, Z. M. Hu, et al., "An experimental study on the percolation characteristics of single phase gas in low and ultra-low permeability sandstone gas reservoirs," Gas Industry, vol. 29, no. 9, pp. 75–77, 2009.
23. S. S. Gao, W. Xiong, and X. G. Liu, "Experimental research status and several novel understandings on gas percolation mechanism in low-permeability sandstone gas reservoirs," Gas Industry, vol. 30, no. 1, pp. 52–55, 2010.
24. G. Y. Zhu, X. G. Liu, S. T. Li, et al., "A study of slippage effect of gas percolation in low permeability gas reservoirs," Gas Industry, vol. 27, no. 5, pp. 44–47, 2007.
25. J. C. Cai, L. J. You, X. Y. Hu, et al., "Prediction of effective permeability in porous media based on spontaneousn imbibition effect," International Journal of Modern Physics C, vol. 23, no. 7, Article ID 1250054, 2012.
26. J. C. Cai, X. Y. Hu, D. C. Standnes, and L. J. You, "An analytical model for spontaneous imbibition in fractal porous media including gravityo," Colloids and Surfaces A, vol. 14, no. 4, pp. 228–233, 2012.

Chapter 9

Methods for Separation, Recycling and Reuse of Biodegradation Products

Ganapati D. Yadav[1] and Jyoti B. Sontakke[1]

[1]Department of Chemical Engineering, Institute of Chemical Technology, Matunga, Mumbai, India

INTRODUCTION

Thousands of chemicals and materials with varied properties and functionalities are manufactured and used for commercial and day-to-day applications, whose ultimate fate in the environment may not be known. During their manufacture and use, these substances are often discharged into the environment through different routes in air, water and land. Creation of tremendous quantities of solid waste of all kind and its effective disposal has posed innumerable problems that need technological breakthroughs. Many of these substances degrade slowly and exert toxic effects on plants and animals, thus causing large scale environmental degradation [1, 2]. Pollution by abandoned plastic articles is also a matter of great concern [3]. Industrial wastewaters associated with the manufacture of organic chemicals are voluminous and characteristically have concentrations ranging from a few ppm to a thousands of ppm. Biodegradation of such dissolved pollutants is an area of immense interest to various sectors. Emission of volatile organic compounds (VOCs) from various sources has detrimental effects on quality of air we breathe and on environmental phenomena. Biodegradation, either aerobic or anaerobic, can be an approach to cleave big molecules through a series of steps in to smaller molecules from a mosaic of chemicals and materials and some of them can be valorized as pollution abatement strategy and source of energy through biogas generation [2]. Biogas can be produced from nearly all kind of biomass, among which the primary agricultural sectors and various organic waste streams can be properly tapped as renewable source of energy. Untreated or poorly managed animal manure is a major source of air and water pollution. Nutrient leaching, mainly nitrogen and phosphorous, ammonia evaporation and pathogen contamination are some of the foremost threats [3]. A conservative estimate is provided by Steinfeld et al. [4] that the animal production sector is responsible for 18% of the overall green house gas emissions, measured in CO_2 equivalent and for 37% of the anthropogenic methane, which has 23 times the global warming potential of CO_2. Furthermore, 65% of anthropogenic nitrous oxide and 64% of anthropogenic ammonia emission originate from the worldwide animal production sector. Biogas production from anaerobic digestion of animal manure and slurries can be harnessed to alleviate greenhouse gas emissions in particularly ammonia and methane [5].

Plastics are bane and benefactor simultaneously. Over 230 million tons of plastic are produced annually. Plastics are used in all walks of life and provide improved insulation, lighter packaging, are found in cars, aeroplanes, railways, phones, computers, medical devices, etc. but appropriate disposal is often not properly addressed. On one hand, plastic waste and disposal is a hotly debated issue globally whereas on the other, it can contribute to reduce the carbon footprint. Many leading European countries recover more than 80% of their used plastics, by adopting an integrated waste and resource management strategy to address each waste stream with the best options [6]. Plastic sorting and separation, recycling, depolymerisation, cracking, and production of fuel are some of the strategies used to abate plastic pollution. Development of biopolymers is pursued vigorously. Biodegradation of plastics by microorganisms and enzymes appears to be the most effective process. When plastics are used as substrates for microorganisms, evaluation of their biodegradability should not only be based on their chemical structure, but also on their physical properties such as melting point, glass transition temperature, crystallinity, storage modulus, etc. [7-11].

This chapter has covered the mechanisms of biodegradation, biodegradation of a variety of industrial chemicals, plastics and other biomass, advances in anaerobic digestion technologies and biogas generation, plastic processing, biopolymer synthesis and degradation. Synthesis of biopolymers is covered. The scope for treating municipal organic solid waste, manure and polymers to generate biogas as a renewable energy option, and also as a pollution abatement strategy is discussed including technological aspects. The synthesis of biohydrogen, bioethanol, biobutanol and other biotransformation leading to valuable chemicals, which also involve breaking down of larger molecules, plastics and biomaterials are not addressed [7,10]. Biorefinery is a concept which is akin to petrorefinery, wherein biomass is converted into useful platform chemicals through extraction, controlled pyrolysis, fermentation, enzyme and chemical catalysis [12].

MECHANISMS OF BIODEGRADATION

Cellulose, lignocellulose and lignin are major sources of plant biomass and are polymeric substances; therefore, their recycling is indispensable for the carbon cycle [13]. Each of these polymer is degraded by a variety of microorganisms which produce scores of enzymes that work in tandem. The diversity of cellulosic and lignocellulosic substrates has contributed to the difficulties found in enzymatic treatment. Fungi are the best-known microorganisms capable of degrading these three polymers. Because the substrates are insoluble, both bacterial and fungal degradation occur exo-cellularly, either in association with the outer cell envelope layer or extra-cellularly. Microorganisms have two types of extracellular enzymatic systems, namely, the hydrolytic system, which produces hydrolases and is responsible for cellulose and hemicellulose degradation; and a unique oxidative and extracellular ligninolytic system, which depolymerizes lignin [13]. The man-made chemicals and materials are comprised of different entities and functional groups which need to be degraded effectively by microorganisms and no single microorganism is obviously capable of doing it [1,14].

Growth and co-metabolism are the two mechanisms of biodegradation. In the case of growth, organic substance is used as the sole source of carbon and energy, which leads to complete degradation (mineralization). Archaebacteria, prokaryotes and eukaryotes (like fungi, algae, yeasts, protozoa) play dominant role in mineralization [7]. On the contrary, co-metabolism encompasses the metabolism of an organic compound in the presence of a growth substrate which is used as the primary carbon and energy source. Thus, biodegradation processes and their rates differ greatly depending on the type of substrate and conditions such as temperature, pH, and aqueous phase solubility, but frequently the major final products of the degradation are carbon dioxide and methane [1, 7, 10].

Growth-Associated Degradation of Aliphatic Compounds

Growth-associated degradation produces CO_2, H_2O, and cell biomass. The cells act as the complex biocatalysts of degradation. Further, cell biomass may be mineralized after exhaustion of the degradable pollutants in a contaminated site. Bulk chemicals like aromatic hydrocarbons such as benzene, toluene, ethylbenzene, xylenes, and naphthalene are widely used as fuels, industrial solvents and feedstock for petrochemical industry. Phenols and chlorophenols are another class of chemicals, employed in a variety of industries. Since all micro-organisms make aromatic compounds such as aromatic amino acids, phenols, or quinines, in large amounts, many microorganisms have evolved catabolic pathways to degrade aromatic compounds. In general, man-made organic chemicals (xenobiotics) can be degraded by microorganisms, when the respective molecules are similar to natural compounds [7,10].

In general, benzene, condensed ring and related compounds are characterized by a higher thermodynamic stability than aliphatic compounds. Benzene oxidation begins with hydroxylation catalyzed by a dioxygenase leading to a diol (Scheme 1) which is then converted to catechol by a dehydrogenase.

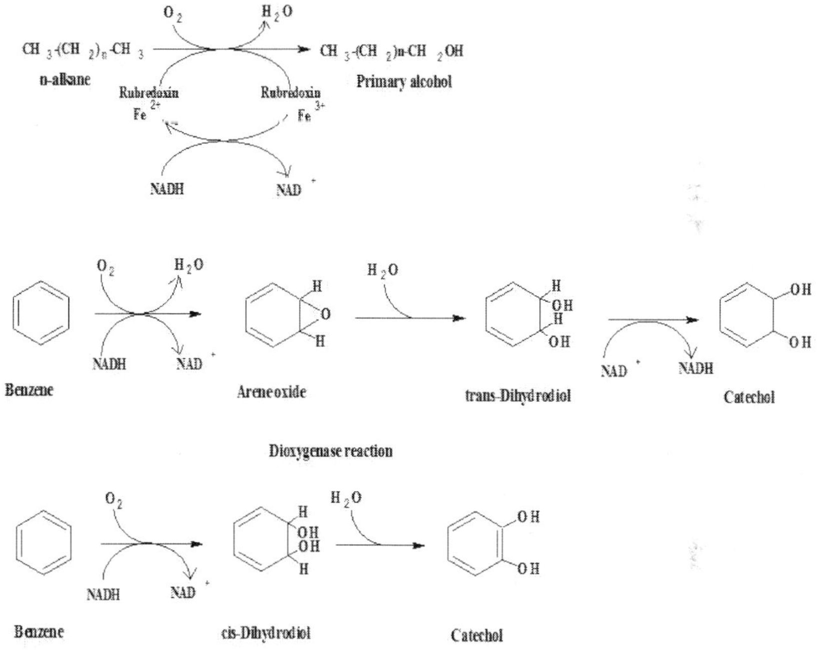

Scheme 1: Monooxygenase and dioxygenase reactions: In this mechanism, monooxygenase initially incorporates one O atom from O_2 into the xenobiotic substrate whereas the other is reduced to H_2O. On the contrary, dioxygenase incorporates both atoms into the substrate [15].

Hydroxylation and dehydrogenation are also common in degradation routes of other aromatic hydrocarbons. The introduction of a substituent group onto the benzene ring renders alternative mechanisms possible to attack side chains or to oxidize the aromatic ring. Many aromatic substrates are degraded by a limited number of reactions such as hydroxylation, oxygenolytic ring cleavage, isomerization, and hydrolysis. The inducible nature of the enzymes and their substrate specificity enable bacteria such as *pseudomonads* and *rhodococci* with a high degradation activity, to acclimatize their metabolism to the effective utilization of substrate mixtures in polluted soils and also to grow at a high rate [10, 15].

Co-Metabolic Degradation of Organo-Pollutants

Co-metabolism is a common phenomenon of microbial activities and the basis of biotransformation used in biotechnology to convert molecules in to useful modified forms. Microorganisms growing on a particular substrate also oxidize a second substrate. The co-substrate is not incorporated, but the product may be available as substrate for other organisms of a mixed culture. The rudiments of co-metabolic transformation are the enzymes of the growing cells and the synthesis of cofactors necessary for enzymatic reactions; for instance, of hydrogen donors (reducing equivalents, NADH) for oxygenases. Several aromatic substrates can be converted enzymatically to natural intermediates of degradation such as catechol and protecatechuate (Scheme 2) [15].

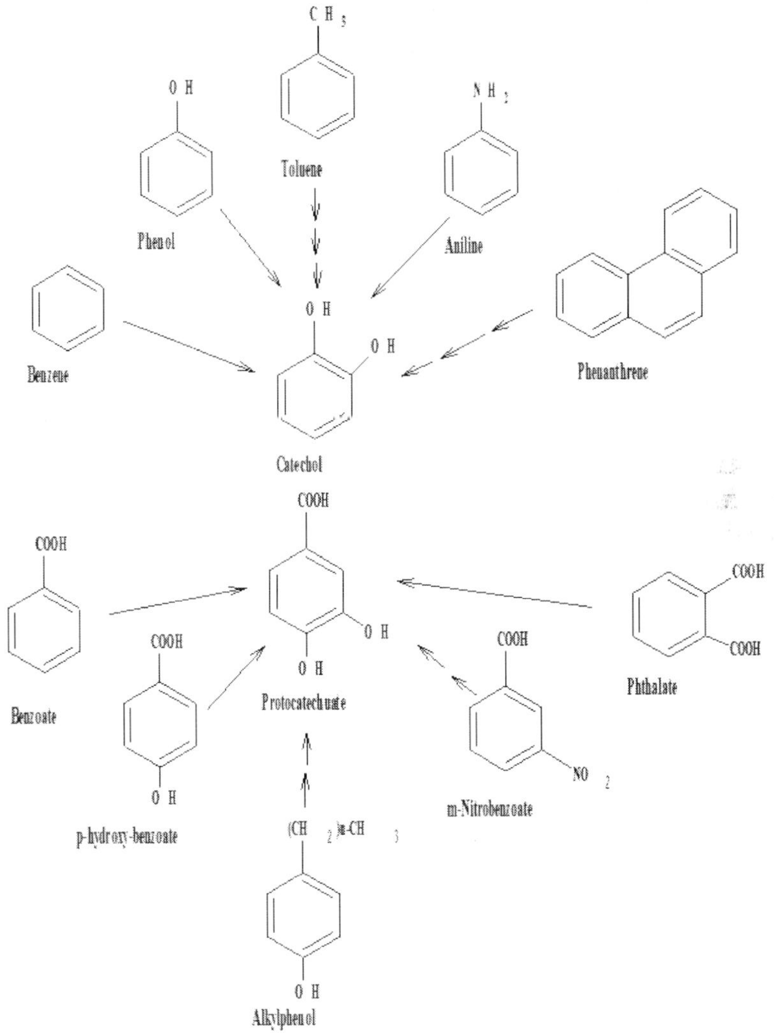

Methods for Separation, Recycling and Reuse of Biodegradation Products

Scheme 2: Degradation of aromatic natural and xenobiotic compounds into two central intermediates, catechol and protocatechuate [15].

Co-metabolism of chloroaromatics is a general activity of bacteria in mixtures of industrial pollutants. The co-metabolic transformation of 2-chlorophenol leads to dead-end metabolites such as 3-chlorocatechol, which may be auto-oxidized or polymerized in soil to humic-like structures. Irreversible binding of dead end metabolites may fulfill the function of detoxification. The accumulation of dead-end products within microbes under selection pressure is the source for the evolution of new catabolic traits. Thus, recalcitrance of organic pollutants increases with increasing halogenation. Substitution of halogen as well as nitro and sulfo groups at the aromatic ring is accomplished by an increasing electrophilicity of the molecule. These compounds resist the electrophilic attack by oxygenases of aerobic bacteria. Compounds that persist under oxic condition are polychlorinated biphenyls (PCBs), chlorinated dioxins and some pesticides like DDT. To overcome the relatively high persistence of halogenated xenobiotics, reductive attack of anaerobic bacteria is of great value. Reductive dehalogenation achieved by anaerobic bacteria is either a gratuitous reaction or a new type of anaerobic respiration. The process reduces the degree of chlorination and, therefore, makes the product more accessible to mineralization by aerobic bacteria [7, 15].

Reductive dehalogenation which is the first step of degradation of PCBs requires anaerobic conditions wherein organic substrates act as electron donors. PCBs accept electrons to allow the anaerobic bacteria to transfer electrons to these compounds. Anaerobic bacteria capable of catalyzing reductive dehalogenation seem to be relatively omnipresent in nature. Most dechlorinating cultures are a mixed consortia. Anaerobic dechlorination is always incomplete and the products are di- and monochlorinated biphenyls. These products can be metabolized further by aerobic microorganisms [2,7,15].

The rates of biodegradability of particular substrate is mainly related to accessibility of the substrate for enzymes and can be enhanced by several means as reviewed by van Lier et al. [16] such as (a) mechanical methods: the disintegration and grinding of solid particles present in sludge: releases cell compounds and creates new surface where biodegradation take place, (b) ultrasonic disintegration, (c) chemical methods: the destruction of complex organic compounds by means of strong, mineral acids or alkalis, (d) thermal pretreatment: thermal hydrolysis is able to split and decompose a significant part of the sludge solid fraction into soluble and less complex molecules, (e) enzymatic and microbial pre-treatment: a very promising method for the future for some specific substrates (e.g. cellulose, lignin etc.),(f) stimulation of anaerobic micro-organisms: some organic compounds (e.g. amino acids, cofactors, cell content) act as a stimulating agent in bacteria growth and methane production. Most of the above methods occur at the pre-methanation step and result in a better supply of methanogenic bacteria by suitable substrates.

AEROBIC BIODEGRADATION

Many microorganisms grow under aerobic conditions. The so-called cellular respiration process (CSP) begins with aerobes which employ oxygen to oxidize substrates such as sugars and fats to derive energy. Before the onset of CSP, glucose molecules are degraded into smaller molecules in the cytoplasm of the aerobes. The smaller molecules then enter a mitochondrion, where aerobic respiration takes place. Oxygen is used to break down small entities into water and carbon dioxide, accompanied by release of energy. Aerobic degradation does not produce foul gases, unlike anaerobic process. The aerobic process leads to a more complete digestion of solid waste reducing build-up by more than 50% in most cases [1, 2, 7]. The major enzymatic reactions of aerobic biodegradation are oxidations catalyzed by oxygenases and peroxidases. Oxygenases are oxido-reductases that incorporate oxygen into the substrate as given in Scheme 1. Degradative organisms need oxygen at two metabolic sites, namely, at the initial attack of the substrate and at the end of the respiratory chain. Higher fungi possess a unique oxidative system for the degradation of lignin based on extracellular ligninolytic peroxidases and laccases [13]. This enzymatic system is important for the co-metabolic degradation of persistent organic pollutants. The predominant bacteria of polluted soils belong to a spectrum of genera and species (Table 1) [15].

The most important classes of organic pollutants in the environment are mineral oil constituents and halogenated petrochemicals, for the biodegradation of which the capacities of aerobic microorganisms are of great consequence. The most rapid and complete degradation of the majority of pollutants is brought about under aerobic conditions and these include petroleum hydrocarbons, chlorinated aliphatics, benzene, toluene, phenol, naphthalene, fluorine, pyrene, chloroanilines, pentachlorophenol and dichlorobenzenes. Many cultures of bacteria grow on these chemicals and are capable of producing enzymes which degrade them into non-toxic species. [7, 15].

Table 1: Predominant bacteria in soil samples polluted with aliphatic and aromatic hydrocarbons, polycyclic aromatic hydrocarbons, and chlorinated compounds [15]

Gram negative bacteria	Gram positive bacteria
Pseudomonas species	Nocardia species
Xanthomonas species	Mycobacteria species
Alcaligenes species	Corynebacterium species
Flavobacterium species	Arthobacter species
Cytophaga group	Bacillus species

There are several essential attributes of aerobic microorganisms degrading organic pollutants amongst which metalobic processes top the list. The chemicals must be accessible to the degrading organisms. For example, hydrocarbons are immiscible in water and their degradation requires the production of biosurfactants in order to have effective biodegradation [14]. The initial intracellular attack of organic pollutants is an oxidative process and therefore, the activation and incorporation of oxygen is the main enzymatic reaction catalyzed by oxygenases and peroxidases. Peripheral degradation pathways convert organic pollutants step by step into intermediates of the central intermediary metabolism, such as the tricarboxylic acid cycle. Biosynthesis of cell biomass from the central precursor metabolites (acetyl-CoA, succinate, pyruvate) is required [14,15]. Sugars needed for various biosyntheses and growth must be synthesized by gluconeogenesis. The predominant degraders of organo-pollutants in the oxic zone of contaminated areas are chemo-organotropic species that are able to use a large number of natural and xenobiotic compounds as carbon sources and electron donors for the generation of energy. Although many bacteria are able to metabolize organic pollutants, a single bacterium does not possess the enzymatic capability to degrade all or even most of the organic pollutants from a heterogeneous mixture originating from particular industries. Thus, mixed microbial communities have the most powerful biodegradative potential. The genetic information of more than one organism is necessary to develop a system which could be used on industrial scale to degrade the complex mixtures of organic compounds present in contaminated areas. The genetic potential and certain environmental factors such as temperature, pH, and available nitrogen and phosphorus sources govern the rate and the extent of degradation [14].

ANAEROBIC BIODEGRADATION

Among biological treatments, anaerobic digestion is frequently the most economical process, due to the high energy recovery linked to the process and its limited environmental impact. Anaerobic biodegradation results when the anaerobic microbes are predominant over the aerobic microbes. Here oxygen does not serve as the final electron acceptor or reactant. Manganese and iron ions, and substances like sulfur, sulfate, nitrate, carbon dioxide, some organic intermediates and pollutants are reduced by electrons originating from oxidation of organic compounds [7]. The common example of anaerobic process is the biodegradable waste in landfill. Paper and other materials degrade more slowly over longer periods of time. Biogas, coming from anaerobic digestion, mainly consists of methane and can be collected efficiently and used for eco-friendly power generation as has been demonstrated on larger scale [3, 16]. Anaerobic digestion is widely used, as part of an integrated waste management system, to treat wastewater sludge and biodegradable waste because it provides volume and mass reduction of the input material. It reduces the emission of landfill gas into the atmosphere [17-20]. Anaerobic digestion is a renewable energy source because the process produces methane and CO_2-rich biogas suitable for energy production helping to replace fossil fuel requirement. Also, the nutrient-rich solids left after digestion can be used as fertilizer [16, 21].

There are four major biological and chemical steps of anaerobic digestion: hydrolysis, acidogenesis, acetogenesis, and methanogenesis [17, 18]. The mechanism commences with bacterial hydrolysis of the organic matter to break down insoluble organic polymers such as carbohydrates and make them available for other bacteria. Acetogenic bacteria convert the sugars and amino acids into carbon dioxide, hydrogen, ammonia, and organic acid. Methanogens then ultimately transform these products in to methane and carbon dioxide [19].

Advances in Anaerobic Digestion Technologies

Thermophilic anaerobic digestion of manure [20] and assessment of biodegradability of macropollutants [21] have demonstrated the prowess of anaerobic digestion which is now a general method used to stabilize municipal wastewater treatment residuals [22, 23]. The so-called phased or staged anaerobic digestion is a recent technology for digestion facilities which include four different configurations of reactors: staged mesophilic digestion, temperature-phased digestion, acid/gas phased digestion, and staged thermophilic digestion [24]. Phased or staged configurations are multiple reactor digestion systems. Phased anaerobic digestion is defined as a digestion system having two or more tanks, each with exclusive operating conditions that support unique biomass populations, which may be acid-forming, methane-forming, thermophilic, or mesophilic organism populations. Effective digestion is achieved by manipulating operational parameters such as solids retention time (SRT) and temperature. Temperature phased digestion system is found better than the other systems during each study phase by having higher volatile solids reduction (VSR), higher methane production, and lower residual biological activity [24,25].

On industrial scale, anaerobic digestion of solid waste is considered as a mature technology [16,26]. Around 60% of the plants are reported in Europe to operate at the mesophilic range (40% thermophilic) with continued increase in capacity over the years in most European countries. Yields from the biomethanization process are very much dependent on operating conditions and the kind of substrate used. Digestion of grey wastes or residual refuse after source separation, has caught attention of industry and some of the solutions considered are landfilling or incineration [23]. However, anaerobic digestion is a better option since it gives number of advantages such as greater flexibility, the possibility of additional material recovery (up to 25%), and a more efficient and ecological energy recovery. In this case the low-calorific organic fraction is digested, the high-calorific fraction is treated thermally and the non-energy fractions can be recovered and reused. It is predicted that this residual refuse will be treated by anaerobic digestion [16, 23].

A very high growth potential is expected for the anaerobic digestion of organic fraction of municipal solid waste (OFMSW). Around 50% of MSW is landfilled, with a content of around 30% of organic fraction (without considering paper and cardboard). The growth potential for this technology is very important to reduce greenhouse gases emission as agreed at the Kyoto Summit [23]. Further, the consolidation of anaerobic digestion as a mainstream technology for the OFMSW should occur since the digested residue can be considered quite stable organic matter with a very slow turnover of several decades given adequate soil conditions. Thus, the natural imbalance in CO_2 can be adjusted by restoring or creating organic rich soil. The removal of CO_2 constitutes an extra benefit that would place anaerobic digestion as one of the most relevant technologies in this field. The degradation of chlorinated compounds need to be examined in greater depth, as anaerobic treatment offers high potential in this area [28].

Several novel reactors with high mass transfer rates, such as fluidized bed reactors, expanded granular sludge bed (EGSB) reactors [29-32], and membrane bioreactors [33] with different configurations have been used, in which hydraulic retention times (HRT) are uncoupled from the solids retention time (SRT) to make anaerobic technology economical alternative for conventional wastewater treatment systems. The upflow anaerobic sludge blanket (UASB) reactors [30] and/or related systems are mostly applied, wherein spontaneous formation of granular conglomerates of the anaerobic organisms occurs, leading to anaerobic sludge with an extremely low sludge volume index and optimal settling properties [21]. Besides, several large scale biogas plants have been built which combine waste from agriculture, industry and households and produce both biogas and a liquid fertiliser which is re-circulated back on agri-land. The combination of anaerobic digestion with other biological or physico-chemical processes has led to the development of optimised processes for the combined removal of organic matter, sulphur and nutrients. In conjunction with anaerobic digestion which removes mainly carbon, other processes are used to remove nitrogen and phosphorus (with oxic phase), which mainly use micro-organisms and also physico-chemical processes. For the treatment of municipal wastewater, the ANANOX process [34] takes advantage of sulphate reduction to sulphide to provide an electron donor for the denitrification process [35-37]. The integration of the nitrogen cycle in anaerobic digestion could be maximised with the application of the ANAMMOX process that makes use of particular micro-organisms that are able to oxidise ammonium to N_2 gas with nitrite as electron acceptor [38, 39].

BIODEGRADATION OF INDUSTRIAL ORGANIC POLLUTANTS

Knowledge of fate of chemicals discharged in the environment, the life cycle analysis and the mechanisms by which they degrade are of great importance in designing biodegradation systems since many of the

Volatile Organic Compounds (VOCS)

There are two classes of VOCs that are responsible for a large number of land and groundwater contamination: (i) petroleum hydrocarbons (PHCs) such as gasoline, diesel, and jet fuel, and (ii) chlorinated hydrocarbon (CHC) solvents such as the dry cleaning agents such as tetrachloroethylene, perchloroethylene (PCE) and the degreasing solvents such as trichloroethylene (TCE), 1,1,1-trichloroethane (TCA), and PCE.

PHCs biodegrade readily under aerobic medium, whereas CHCs characteristically biodegrade much more slowly and under anaerobic conditions [43]. Because PHC biodegradation is relatively rapid when oxygen is present, aerobic biodegradation can usually limit the concentration and subsurface migration of petroleum vapours in unsaturated soils. Further, CHC biodegradation can produce toxic moieties, such as dichloroethylene and vinyl chloride, while petroleum degradation usually produces carbon dioxide, water, and sometimes methane or other simple hydrocarbons. A second primary difference is density of pollutant. PHC liquids are lighter than water and immiscible. PHCs can float on the groundwater surface (water table), whereas chlorinated solvents being heavier than water sink through the groundwater column to the bottom of the aquifer. These major differences in biodegradability and density lead to very different subsurface behaviour that often reduces the potential for human exposure.

Petroleum Hydrocarbons (PHCS)

It is known that microorganisms capable of aerobically degrading PHCs are present in nearly all subsurface soil environments [44-49]. Effective aerobic biodegradation of PHCs hinges on the soil having adequate oxygen and water content to provide a habitat for sufficient populations of active microorganisms. If oxygen is present, these organisms will generally consume available PHCs. Furthermore, aerobic biodegradation of petroleum compounds can occur relatively quickly, with degradation half lives as short as hours or days under some conditions [50]. Some petroleum compounds can also biodegrade under anaerobic conditions; however, above the water table, where oxygen is usually available in the soil zone, this process is insignificant and often much slower than aerobic biodegradation. Aerobic biodegradation consumes oxygen and generates carbon dioxide and water. This leads to a characteristic vertical concentration profile in the unsaturated zone in which oxygen concentrations decrease with depth and VOCs including PHCs and methane from anaerobic biodegradation and carbon dioxide concentrations increase with depth [51, 52].

Chlorinated Hydrocarbon (CHC) Solvents

Chlorinated solvents such as tetrachloroethylene (TCE), 1,1,2,2-tetrachloroethane, carbon tetrachloride, and chloroform are released as waste products by spills, land-filling, and discharge to sewers during manufacture and their use as solvents in a variety of cleaning processes or as vehicles for solid slurries. TCE is a major pollutant of the industry. It is biodegraded under anaerobic conditions through hydrogenolysis that sequentially produces isomers of 1, 2-dichloroethylene (1, 2-DCE), vinyl chloride (VC), and ethylene. Some labs have also reported ethane [53, 54], methane [55], and carbon dioxide [56] as degradation products.

In addition to anaerobic degradation through reductive dechlorination (hydrogenolysis), TCE and other chlorinated VOCs can be susceptible to co-metabolic oxidation by aerobic microorganisms that have oxygenases with broad substrate specificity. Methanotrophs are microorganisms that primarily oxidize methane for energy and growth using methane monooxygenase (MMO) enzymes and are a group of aerobic bacteria transform TCE through co-metabolic oxidation [57-59]. In contrast to reductive dechlorination, where the degradation rate generally decreases as the degree of chlorination of the aliphatic hydrocarbon decreases, the less-chlorinated VOCs such as 1,2-DCE and VC are more straightfowardly and quickly degraded through aerobic oxidation reactions than the higher chlorinated compounds such as TCE [60]. Methane-oxidizing bacteria are known to convert TCE to its epoxide, which then breaks down immediately in water to form dichloroacetic acid, glyoxylic acid, or one-carbon compounds such as formate or CO. The two carbon acids accumulate in the water phase, while formate and CO are further oxidized by methanotrophic bacteria to CO_2. Hence, coupling of anaerobic and aerobic degradation processes has been recommended as the best possible bioremediation method for chlorinated VOCs such as TCE [60-62].

Quinoline

Quinoline occurs commonly in coal tar, oil shale, and petroleum, and is used as an intermediate and solvent in many industries [63, 64]. Due to its toxicity and repulsive odor, quinoline-containing waste is detrimental to human health and environmental quality. The study of quinoline- degrading bacteria not only helps to reveal the metabolic mechanism of quinoline, but also benefits the bio-treatment of quinoline-containing wastewater. Although different genera of bacteria may produce different intermediates, almost all of them transform quinoline into 2-hydroxyquinoline in the first step [63, 65]. A quinoline-degrading bacteria strain, *Pseudomonas* sp. BW003, was isolated from the activated sludge in a coking wastewater treatment plant. *Pseudomonas* strains degrade quinoline via the 2-hydroxyquinoline and 2,8-hydroquinoline pathway, and then transform 2,8-hydroquinoline into 8-hydrocumarin, which is then transformed into 2,3-dihydroxyphenylpropionic acid, and finally to CO_2 and H_2O (Scheme 3) [66-69]. Quinoline-N is transformed into ammonia-N, as reported in few genera of bacteria. Thus, quinoline pollution can be eliminated by applying such degrading bacteria in the treatment with bio-augmentation [70-72].

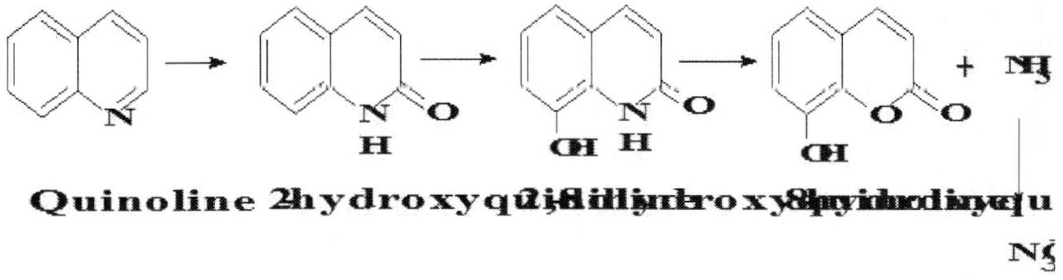

Quinoline → 2-hydroxyquinoline → 2,8-dihydroxyquinoline → 8-hydroxycoumarin + NH_3

Scheme 3: Degradation products of quinoline [63].

Phenols

Phenols are harmful to organisms at low concentrations and classified as hazardous pollutants because of their potential to harm human health. They exist in different concentrations in wastewaters originated from coking, synthetic rubber, plastics, paper, oil, gasoline, etc. Biological treatment, activated carbon adsorption and solvent extraction are some of the most widely used methods for removing phenol and family compounds from wastewaters [73-76]. Biological treatment is economical, practical, promising and versatile approach for it leads to complete mineralization of phenol. Many aerobic bacteria are capable of using phenol as the sole source of carbon and energy [77]. In recent years, the strain of *Pseudomonas putida* has been the most widely used to degrade phenol. Under aerobic conditions, phenol may be converted by the bacterial biomass to CO_2; other intermediates such as benzoate, catechol, *cis*-cis-muconate, β-ketoadipate, succinate and acetate are formed during the biodegradation process [77, 78]. *p*-Nitrophenol (PNP) is one of the most widely used nitrophenolic compounds in industry and finds important applications in agriculture, polymers, pigment and pharmaceutical industries. However, PNP is highly toxic for both the environment and humans and its efficient removal from the environment is required. Hydroquinone (HQ), 4-nitrocatechol (4-NC) and 1,2,4-benzenetriol (1,2,4-BT) are the metabolic intermediates of the PNP biodegradation [80,81].

Chlorinated phenols are common and encountered even in relatively pristine environments [82,83]. These compounds are formed during the bleaching of pulp with chlorine [82-84]. As the pulp accounts only for about 40-45% of the original weight of the wood, these effluents are heavily loaded with organics [85]. Chlorophenols are also used as fungicides and may be formed from hydrolysis of chlorinated phenoxyacetic acid herbicides. Chlorophenols, part of the *adsorbable organic halides*(AOX), are present in bleaching effluents at concentrations ranging from 0.1 to 2.6 ppm [86]. Aqueous effluents from industrial operations such as polymeric resin production, oil refining and coking plants also contain chlorophenolic compounds. Pentachlorophenol (PCP) is the second most heavily used pesticide in the US. As compared to phenol, chlorophenolic compounds are more persistent in the environment. Toxicity and bioaccumulative potential of chlorophenols increases with the degree of chlorination and with chlorophenol lipophilicity. Haloaromatic compounds are degraded via the formation of halocatechols as intermediates which are subsequently cleaved by dioxygenases, by the mechanism delineated earlier. Dehalogenation then occurs by the elimination of the hydrogen halide, with subsequent double bond formation on the aliphatic intermediate [87]. In anaerobic environments, the biodegradation of chlorinated aromatics takes place through reductive dehalogenation leading to the formation of less toxic and more biodegradable compounds. Reductive dechlorination of 2,4-dichlorophenol is followed by carboxylation, ring fission and acetogenesis, and methanogenesis which finally led to the complete mineralization of 2,4-DCP, which is also biodegraded to 4-chlorophenol in anaerobic sediments. Similarly, biodegradation of PCP under anaerobic conditions occurs through reductive dechlorination [88].

Fluoro Benzenes

Toluene degrading enzymes can transform many 3-fluoro-substituted benzenes to the corresponding 2,3-catechols with the concomitant release of inorganic fluoride. The substrates that induce 2,3-dioxygenase are 3-fluorotoluene, 3-fluorotrifluorotoluene, 3-flurohalobenzene, 3-fluoronisole, and 3-fluorobenzonitrile. While 3-fluorotoluene and 3-fluoronisole produce only deflorinated catechols, other substrates led to catechol products both with and without the toluene substituent [89].

Polycyclic Aromatic Hydrocarbons (PCAHS)

PCAHs are toxic, mutagenic and resist biodegradation [90]. Many strategies have been developed to treat them, including volatilization, photooxidation, chemical oxidation, bioaccumulation, and adsorption on soil particles [91]. Soil clean-up may be achieved using different remediation technologies, among which bioremediation is an effective and low-cost alternative that has garnered widespread use [92]. Two processes have been found to increase the activity of microorganisms during bioremediation: bio-stimulation and bio-augmentation. Bio-stimulation involves the addition of nutrients and/or a terminal electron acceptor to increase the meager activity of indigenous microbial populations. Bio-augmentation involves the addition of external microbial strains (indigenous or exogenous) which have the ability to degrade the desired toxic compounds [93]. The added specific PCAHs degrading strain, which has a competitive capacity to become dominant species with indigenous microbial strains or grow simultaneously with indigenous microbial strains, may greatly enhance the rate of PCAHs biodegradation [94,95]. The ability to degrade PCAHs depends on the complexity of their structure and the extent of enzymatic adaptation by bacteria. In general PCAHs with 2 or 3 aromatic rings are readily degraded, but those with 4 or more are usually recalcitrant and genotoxic. Such examples of PCAHs are acenaphthene, acenaphthylene, anthracene, naphthalene, fluorene, phenanthrene, chrysene, pyrene, etc. The major metabolites are 4-phenanthroic acid and 4-hydroxyperinapthenone. Cinnamic and phthalic acids are ring fission products [96].

Naphthalene is carcinogenic and persistent organic pollutant [97]. Bacteria such as *Pseudomonas putida*, *Rhodococcus opacus*, *Mycobacterium* sp., *Nocardia otitidiscaviarum*, and *Bacillus pumilus* are known to biodegrade naphthalene [98-102]. Some metabolites of naphthalene, such as salicylic acid, 1-naphthol and *o*-phthalic acid could be degraded and support cell growth (Scheme 4). Phenanthrene was used as a model compound for PCAH degradation which shows 1-hydroxy 2-naphthoic acid (1H2NA) as intermediate biodegradation product [103].

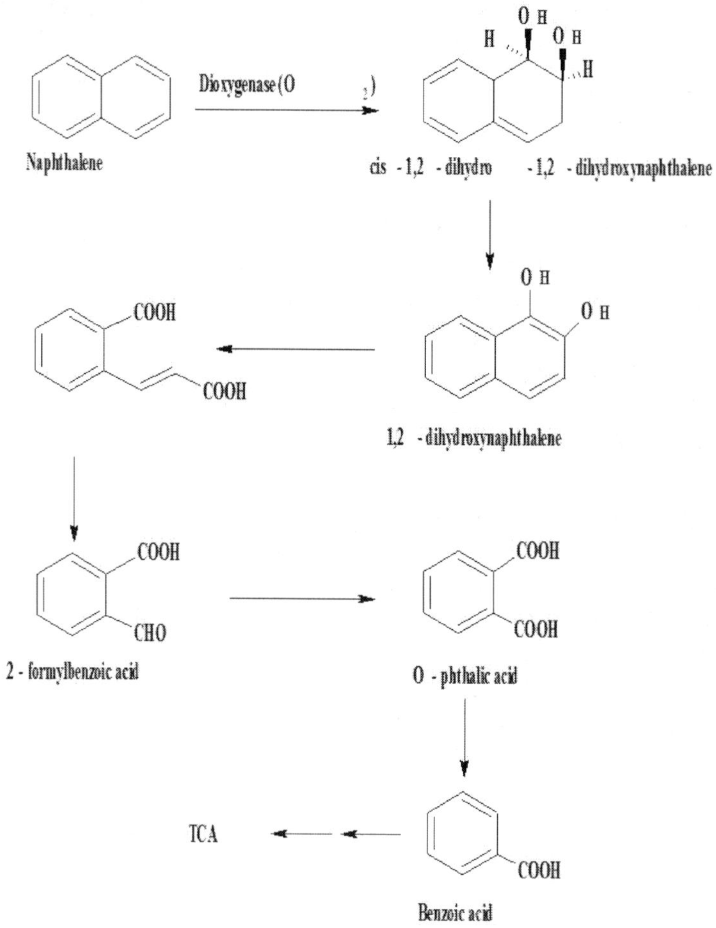

Scheme 4: Proposed pathway for the degradation of naphthalene [103]

Plasticizers

Plasticizers are polymeric additives, used to impart flexibility to polymer materials. The biodegradation of some plasticizers can lead to the formation of metabolites with increased persistence and toxicity relative to the original compounds [104-106]. Use of plasticizers has grown considerably, both with respect to product variety and production volume [107]. Phthalates are the most widely used plasticizers. Presence of phthalates and their metabolites in rats, mice, human plasma and liver are related to adverse health effects such as endocrine disruption and peroxisome proliferation [108,109]. The high production volumes of phthalates and their incomplete biodegradation have led to the presence of these compounds and a number of toxic and stable metabolites in surface waters, groundwater, air, soil and tissue of living organisms [104, 110-113]. Such findings have led to stricter environmental regulations and consequently have broadened the criteria used to evaluate plasticizers. Consequently, dibenzoates have been approved by the European Chemical Agency as alternatives to phthalates [114]. However, the degradation of dipropylene glycol dibenzoate (D(PG)DB) and diethylene glycol dibenzoate (D(EG)DB) by common soil microorganisms such as *Rhodotorula rubra* and *Rhodococcus rhodochrous* can lead to the formation and accumulation of monobenzoate metabolites [115,116] that exhibit high acute toxicity [115]. Other compounds including 1,5-pentanediol and 1,6- hexanediol dibenzoates were reported to produce less stable metabolites and have also been tested as potential alternatives to commercial dibenzoate plasticizers [116-118]. Scheme 5 shows the biodegradation products of dibenzoates by *R. Rhodochrous*, which include 2-[2-(benzoyloxy)propoxy] propanoic acid, 1,3-propanediol monobenzoate and 3-(benzoyloxy) propanoic acid [119].

Plastics

Over the years, plastics have brought economic, environmental and social advantages. Today's material world uses tremendous quantities of plastics of all hue and origins. However, their wide spread use has also increased plastic waste, which brings its own economic, environmental and social problems. The redesign of plastic products, whether individual polymer or product structure, could help alleviate some of the problems associated with plastic waste. Redesign could have an impact at all levels of the hierarchy established by the European Waste Framework Directive: prevention, re-use, recycle, recovery and disposal [120].

Polyethylene, polypropylene and polystyrene, and water-soluble polymers, such as polyacrylamide, polyvinyl alcohol and polyacrylic acid are bulk polymers used in a variety of industries and products. Some of the plastics are not biodegradable and deleterious to the environment due to their accumulation. Plastics can be disposed of by incineration or recycling, but incineration is very difficult, dangerous and expensive whereas recycling is a long process and not very efficient. Some plastics still cannot be recycled or incinerated due to pigments, coatings and other additives during manufacture of materials. Making biodegradable and ecofriendly plastics will avoid accumulation, recycling and incineration [121].

Polyvinyl Alcohol

Polyvinyl alcohol (PVA) is water-soluble but also has thermoplasticity. In addition to its use as a water-soluble polymer, for instance, as a substituent for starch in industrial processes, it can also be molded in various shapes, such as containers and films. PVA can therefore be used to make water-soluble and biodegradable carriers, which may be useful in the manufacture of delivery systems for chemicals such as fertilizers, pesticides, and herbicides. Among the vinyl polymers produced industrially, PVA is the only one known to be mineralized by microorganisms [122]. Extensive use of PVA, in textile and paper industries generates considerable amount of contaminated wastewaters [121]. In aqueous solution, the biodegradation mechanism of PVA involves the random endocleavage of the polymer chains. The initial step is associated with the specific oxidation of methane-carbon bearing the hydroxyl group, as mediated by oxidase and dehydrogenase type enzymes, to give β-hydroxyketone as well as 1,3-diketone moieties. The latter groups are able to facilitate the carbon-carbon bond cleavage as promoted by specific β-diketone hydrolase, leading to the formation of carboxylic and methyl ketone end groups [123,124]. Most of the PVA-degrading microorganisms are aerobic bacteria belonging to *Pseudomonas*, *Alcaligenes*, and *Bacillus* genus. A very moderate PVA biodegradation was reported [125-128].

Polyhydroxyalkanoates

Polyhydroxyalkanoates (PHAs) are degraded to CO_2 and water in aerobic conditions and methane in anaerobic conditions by microbes found in soil, water and other various natural habitats. PHAs are the only proposed replacement polymers that are completely biodegradable [129]. The structures of these polymers have a very similar structure of the petroleum-derived thermoplastics [130].

Scheme 5: Proposed biodegradation pathways of diethylene glycol dibenzoate and 1, 3-propanediol dibenzoate [116].

PHAs also possess similar physical properties as plastics including the ability to be molded, made into films, and also into fibers. Efforts are underway to identify bacteria, which produce various kinds of PHAs [129] as well as the production of these polyesters or create certain kinds of PHAs by changing the kind of bacteria [130] and/or the substrates given to the bacteria and genetically enhancing bacteria [131].

PROSPECTIVE OF ANAEROBIC DIGESTION AND BIOGAS ENERGY

The foregoing analysis shows that anaerobic digestion technologies have matured so far to treat several organic micro-pollutants, halogenated compounds, substituted aromatics, azo-linkages, nitro-aromatics and the like in industrial effluents and also for municipal effluents containing industrial loads. A very high growth potential is envisaged for the anaerobic digestion of organic fraction of municipal solid waste [27]. Novel reactor and control systems ought to be developed for different purposes depending on the source of pollutants or biomass. Anaerobic digestion of sewage sludge followed by recycling on agricultural land is currently the largest world-wide application of anaerobic processes. Treatment of sludge and slurries targeted at the production of safe end products can be tackled with niche anaerobic technologies [16]. It is predicted that major future process developments will come from the deployment of pre- and post treatment processes, including physical, chemical and biological processes, for the reclamation of the products from the wastewater treatment system. Wastewater treatment for reuse will be effective if anaerobic digestion is adopted for mineralizing organic matter. Hence, anaerobic digestion has the potential to play a major role in closing water, raw materials, and nutrient cycles in industrial processes [37]. Further development is required on the community on-site treatment of domestic sewage under a wide range of conditions, opting for the reuse of the treated water in agriculture and making use of the mineralized nutrients for fertilization purposes. An upstream integration of the anaerobic process with industrial primary production processes under extreme conditions of temperature, pH, salinity, toxic and recalcitrant compounds, and variable load is envisaged in future [39].

There is an emphasis worldwide on renewable energy system among which biogas produced from any biological feedstocks including primary agricultural sectors and from various organic waste streams will come in to prominence in near future [22]. It is estimated [3] that at least 25% of all bioenergy in the future can originate from biogas, produced from wet organic materials like animal manure, slurries from cattle and pig production units as well as from poultry, fish and fur, whole crop silages, wet food and feed wastes, etc. Anaerobic digestion of animal manure offers several environmental, agricultural and socio-economic benefits throughout such as improved fertilizer quality of manure, considerable reduction of odors and inactivation of pathogens and more importantly production of biogas production, as clean, renewable fuel, for multiple utilizations [16]. This biogas can be upgraded to natural gas to mix with the existing natural gas grid which will be cost effective. The potential development of biogas from manure co-digestion includes the use of new feedstock types such as by-products from food processing industries, bio-slurries from biofuels processing industries as well as the biological degradation of toxic organic wastes from pharmaceutical industries, etc. [3,16,22]. This will also call for better reactor systems and careful process control to increase the biogas yield, which will be more attractive if coupled with less capital and operating costs. Integration of biogas production into the national energy grids will eventually be commercially viable since the biogas from anaerobic co-digestion of animal manure and suitable organic wastes would overcome the major environmental and veterinary problems of the animal production and organic waste disposal.

PLASTIC WASTE SEPARATION, REPROCESSING AND RECYCLE

In 2009, around 230 million tonnes of plastic was produced; ~25 % which was used in the European Union [131]. About 50 % plastic is used for single-use disposable applications, such as packaging, agricultural films and disposable consumer items [132]. Although plastics consume approximately 8 % world oil production: 4 % as raw material for plastics and 3-4 % as energy for manufacture [132], supplies are being depleted. Bioplastics make up only 0.1 to 0.2 % total plastics [115]. It is estimated that plastics reduce 600 to 1300 million tonnes of CO_2 through the replacement of less efficient materials, lighter and fuel efficient vehicles, housing sector, contribution to insulation, preservation of food, packaging, use in wind power rotors and solar panels [133]. However, plastic littering and pollution of land and sea have been matters of great concern which should be strategically and technologically solved. Plastics recovery, in addition to increased diversion from disposal, results in significant energy savings (~50-75

MBtu/ton of material recycled) compared with the production of virgin materials, which also leads to reductions in greenhouse gas emissions due to avoided fuel use. Limiting the plastics that enter landfills can lower the costs associated with waste disposal. It is believed that the recycled plastic will fetch as much as 70 % of the typical price for virgin plastics [136].

Waste Reduction Hierarchy

The motto of waste reduction by plastics is by following the principles of (i) prevention, (ii) reuse (iii) recycle, (iv) recovery, and (v) disposal [119].

- *Prevention* – Using minimum and as less types of plastic in the product by clever product redesign.
- *Reuse* – Products could be designed for reuse by facilitating the dismantling of products and replacement of parts. This could involve standardizing parts across manufacturers [137].
- *Recycle* – Some types of plastics are easier to recycle than others, for example polyethylene terephthalate (PET). By using fewer types and colors (or colorless) of plastics the recycling process becomes easier. The use of "intelligent" or smart polymers that undergo changes under certain conditions could also reduce disassembly time [138]. For example, a polymer that changes shape when subject to magnetic or electric fields could aid the disassembly of electronic goods.
- Recovery – Energy can be recovered from plastics in waste-to-energy plants. By designing products to consider the possibility of energy recovery, plastic may have a greater end-of-life use.
- *Disposal* – Biodegradable plastics are less persistent in the environment than traditional plastics, but need specific and suitable end-of-life treatment.

Bioplastics

Since disposal is one of the important aspect, bioplastics are being favored. There are three main categories of bio-based plastics: (i) Natural polymers from renewable sources, such as cellulose, starch and plant-based proteins, (ii) Polymers synthesised from monomers derived from renewable resources. For example, polylactic acid (PLA) is produced by the fermentation of starch, corn or sugar, (iii) Polymers produced by microorganisms. For example, PHA (polyhydroxyalkanoate) is produced by bacteria through fermentation of sugar or lipids [139].

Biodegradable plastics are not by definition bio-based and bio-based plastics are not always biodegradable, although some fall into both categories, such as PHAs. The term *bioplastics* is often used to refer to both bio-based and biodegradable plastics. The main applications of bioplastics are disposable plastic bags, packaging and loose fill packaging (beads and chips), dishes and cutlery, electronic casings and car components. However, bioplastics cannot substitute all types of plastic; particularly certain types of food packaging that require gas permeability [135]. Development of novel biodegradable plastic is a solution for the plastic disposal problem since plastics are immiscible in water and are thermo-elastic polymeric materials. Biodegradability of plastics is governed by both their chemical and physical properties. Other factors affecting degradability are the forces associated with covalent bonds of polymer molecules, hydrogen bonds, van der Waals forces, coulombic forces, etc. Enzymatic degradation is an effective way. Lipase and esterase can hydrolyze fatty acid esters, triglycerides and aliphatic polyesters. These lipolytic enzymes have an important role in the degradation of natural aliphatic polyesters such as cutin, suberin and esteroid in the natural environment and animal digestive tract.

As stated earlier, biodegradable plastics decompose in the natural environment from the action of bacteria. Biodegradation of plastics can be achieved through the action of micro-bacteria and fungi in the environment to metabolize the molecular structure of plastic films to produce an inert humus-like material that is less harmful to the environment, along with water, carbon dioxide and/or methane. They may be composed of either bioplastics or petro-plastics. The use of bio-active compounds compounded with swelling agents ensures that, when combined with heat and moisture, they expand the plastic's molecular structure and allow the bio-active compounds to metabolize and neutralize the plastic [140]. Compostable plastics are biodegradable and meet certain criteria, such as rate of biodegradation and impact on the environment. Degradable plastics include biodegradable and compostable plastics, but also plastics that degrade by chemical and physical processes such as the action of sunlight. Purely biodegradable plastics are different from oxy-biodegradable plastics, which contain small amounts of metal salts to speed up degradation. It has been suggested that this process be called "oxo-fragmentation" to avoid confusion [139,140].

It is possible to produce polymers biologically, e.g., PHA grown in genetically modified corn plant leaves, PLA (polylactic acid) produced by the fermentation of sugars extracted from plants, PHA produced by bacteria. Bioplastics could also help alleviate climate change by reducing the use of petroleum for the manufacture of traditional plastics. It is claimed that CO_2 emissions released at the end-of-life of bio-based plastics are offset by absorption of CO_2 during the growth of plants for their production [141].

Sorting Plastic Materials

The technical difficulties and high cost associated with separating plastics have limited recycling in the past. Consumer goods often contain as many as 20 different types of plastic as well as non-plastic materials such as wood, rubber, glass, and fibers. There is upsurge of new products and pigment types, which can pose a challenge to the recycling infrastructure. Consequently, the cost of producing virgin materials is often less than the cost of collecting and processing post-consumer plastics. Used plastic material will contain more than one base polymer, and resins with a variety of additives, including coloring agents and thus technologies for cleaning and separating the materials are an important part of most plastics recycling systems. A particular concern for recycled plastics is their use as food containers requiring stringent regulations to avoid contamination [140].

Separation of different types of polymers from each other is many times a desired part of plastics recycling processes which are classified as macrosorting, microsorting, or molecular sorting.

Macrosorting

Macrosorting involves the sorting of whole or nearly whole objects such as separation of PVC bottles or caps from PET bottles, separation of polyester carpet from nylon carpet, and sorting of automobile components by resin type. Various devices are now commercially available to separate plastics by resin type, which typically rely on differences in the absorption or transmission of certain wavelengths of electromagnetic radiation, or color or resin type. Particularly for recycling of appliances, carpet, and automobile plastics, several IR spectra based equipment are used [135].

Microsorting

Microsorting is a size-reduction process to reduce the plastic material in to small pieces which is then separated by resin type or color; for instance, separation of high-density polyethylene (HDPE) base cups from PET soft drink bottles using a sink-float tank. More modern separation processes, such as the use of hydrocyclones, also rely primarily on differences in the density of the materials for the separation. A number of other characteristics have also been used as the basis for microsorting systems, including differences in melting point and in triboelectric behavior. In many of these systems, proper control over the size of the plastic flakes is important in being able to reliably separate the resins. Some systems rely on differences in the grinding behavior of the plastics combined with sieving or other size-based separation mechanisms for sorting. Sometimes cryogenic grinding is used to facilitate grinding and to generate size differences [135].

Three new separation technologies, developed by MBA Polymers, Argonne National Laboratory, and Recovery Plastics International (RPI), could break down these barriers and increase plastics recycling [138].

Automated Separation

According to the process developed by MBA Polymers, plastic scraps from computers and other electronics are first ground into small pieces. Magnets and eddy-current separators then extract ferrous and non-ferrous metals. Paper and other lighter materials are removed with jets of air. Finally, a proprietary sorting, cleaning, and testing process involving various technologies, allows the separation of different types of plastics and compound them into pelletized products comparable to virgin plastics [138].

Froth Flotation

Argonne National Laboratory (ANL) has developed a process to separate acrylonitrile-butadiene styrene (ABS) and high-impact polystyrene (HIPS) from recovered automobiles and appliances. The froth flotation process separates two or more equivalent-density plastics by modifying the effective density of the plastics. There is a careful control of the chemistry of the aqueous "froth" so that small gas bubbles adhere to the solid surface and facilitate the plastic to float to the top [135].

Skin Flotation

Recovery Plastics International (RPI) has developed an automated process capable of recovering up to 80 % plastics found in automobile shredder residue (ASR). The process starts with the separation of light lint materials, followed by the removal of rocks and metals, granulation, washing, and, finally, automated skin flotation separation. This final step adds a skin of plasticizer to make the surface of the targeted plastic hydrophobic. Air bubbles then can attach to the plastic, allowing it to float above denser, uncoated pieces. It is estimated this new skin flotation technology could be used to treat about one-third of the estimated 7 million tons of ASR disposed off each year [141].

Molecular Sorting

Molecular sorting deals with sorting of materials whose physical form has been completely disrupted, such as by dissolving the plastics in solvents using either a single solvent at several temperatures or mixed solvents, followed by reprecipitation. There is a need to control emissions and to recover the solvents, without any residual solvent in the recovered polymer to avoid leaching in stored material. There are at present no commercial systems using this approach. Some research effort has focused on facilitating plastics separation by incorporating chemical tracers into plastics, particularly packaging materials, so that they can be more easily identified and separated.

It has become obvious that many of the difficulties of recycling plastics are related to difficulties in separating plastics from other wastes and in sorting plastics by resin type. Design of products can do a lot to either aggravate or minimize these difficulties [134,135]. The concept of green product embeds recycling at the design stage itself.

Plastic Reprocessing and Recycling

For plastics recycling to be effective, it is necessary to have (i) a system for collecting the targeted materials, (ii) a facility capable of processing the collected recyclables into a form which can be utilized to make a new product and, (iii) new products made in whole or part from the recycled material must be manufactured and sold.

Processing of recyclable plastics is necessary to transform the collected materials into raw materials for the manufacture of new products. Three general categories of processing can be identified: (1) physical recycling, (2) chemical recycling, and (3) thermal recycling, wherein the particulars of the processing are often specific to a given plastic or product.

Physical Processes

Physical recycling, the most popular option, covers size and shape alteration, removing contaminants, blending in additives if desired, and similar approaches that change the appearance of the recycled material, but do not alter its basic chemical structure. Plastic containers, for example, are processed including grinding, air classification to remove light contaminants, washing, gravity-separation, screening, rinsing, drying, and often melting and pelletization, accompanied by addition of colorants, heat stabilizers, or other ingredients, depending on type of plastic [132].

Chemical Reactions

Chemical recycling of plastics deals with chemical reactions using catalysis or solvents such as methanol, glycols or water leading to depolymerization or breaking polymers into monomers or useful chemicals, or fuels [134]. The products of the reaction then can be separated and reused to produce either the same or a related polymer. An example is the glycolysis process sometimes used to recycle polyethylene terephthalate (PET), in which the PET is broken down into monomers, crystallized, and repolymerized. Condensation polymers, such as PET, nylon, and polyurethane, typically much more amenable to chemical recycling than are addition polymers such as polyolefins, polystyrene, and polyvinyl chloride (PVC). Most commercial processes for depolymerization and repolymerization are restricted to a single polymer, which is usually PET, nylon 6, or polyurethane. Methanolysis is another common reaction using methanol [134].

Thermal Cracking

Thermal cracking or recycling also involves cracking of the chemical structure of the polymer using heat in the absence of sufficient oxygen for combustion. At these elevated temperatures, the polymeric structure breaks down. Thermal recycling can be applied to all types of polymers. However, the typical yield is a complex mixture of products, even when the feedstock is a single polymer resin. If reasonably pure compounds can be recovered, products of thermal recycling can be used as feedstock for new materials. When the products are a complex mixture which is difficult to separate, they are most often used as fuel. There are relatively few commercial operations today which involve thermal recycling of plastics. Some European nations have such feedstock recycling facilities. Many plastic resin companies use fluidized bed cracking to produce a waxlike material from mixed plastic waste [134-136, 139]. The product, when mixed with naptha, can be used as a raw material in a cracker or refinery to produce feedstocks such as ethylene and propylene. In certain case, syn gas can be produced and used in Fisher-Tropsch synthesis to produce valuable chemicals.

In landfill, both synthetic and naturally occurring polymers do not get the necessary exposure to UV and microbes to degrade. The discarded plastics occupy space and none of the energy put into making them is being reclaimed. Reclaiming the energy stored in the polymers can be done through incineration, but this can cause environmental damage by release of toxic gases into the atmosphere. Therefore, recycling is a viable alternative in getting back some of this energy in the case of some polymers. With ever increasing petroleum prices, it would be financially viable to recycle polymers rather than produce them from raw materials [141].

CONCLUSIONS

The modern society needs thousands of chemicals and materials of all sorts which are produced annually and used in all sectors of economy. However, their fate in the environment is of great concern since some are toxic, recalcitrant and bioacumulating and hence their discharge into the environment must be regulated. Better understanding of the mechanism of biodegradation has a high ecological significance that depends on indigenous microorganisms to transform or mineralize the organic contaminants. Thus, biodegradation processes differ greatly depending on conditions, but frequently the main final products of the degradation are carbon dioxide and/or methane. Microorganisms have enzyme systems to degrade and utilize different hydrocarbons as a source of carbon and energy. Slow and partial biodegradation of chlorophenolic compounds under aerobic as well as anaerobic natural environment has been observed. Aerobic degradation takes place via formation of catechols while anaerobic degradation occurs via reductive dechlorination. Acclimatization of biomass to chlorophenols markedly enhances their ability to degrade such compounds, both by reducing the initial lag phase as well as by countering biomass inhibition. Aerobic processes as well as anaerobic processes partially remove chlorophenols. However, enhanced removal efficiency can be obtained by operating anaerobic and aerobic treatment processes in combination. Thus microbial degradation can be a key component for clean-up strategy of organopollutants and plastics.

Renewable energy system among which biogas produced from biological feedstocks will play a major role in energy sector. Anaerobic digestion of animal manure, slurries from cattle and pig production units as well as from poultry, fish and fur, whole crop silages, wet food and feed wastes, etc offers several environmental, agricultural and socio-economic benefits by improved fertilizer quality of manure, considerable reduction of odors, inactivation of pathogens and production of biogas production, as clean and renewable fuel. This biogas can be upgraded to natural gas to inject in to the existing natural gas grid which will be cost effective. Biogas from anaerobic co-digestion of animal manure and suitable organic wastes would overcome the major environmental and veterinary problems of the animal production and organic waste disposal.

The recycling of plastics is environmentally beneficial because plastics reduce millions of tonnes of CO_2 emissions through the replacement of less efficient materials, development of lighter and fuel efficient transport systems, housing material, energy saving insulation, food preservation and storage, energy efficient packaging, use in wind power rotors and solar panels. Processing of recyclable plastics is necessary to transform the collected materials into raw materials for the manufacture of new products. Bioplastics offer a very good solution to environmentally deleterious materials. Biodegradation of plastics can be achieved through the action of micro-bacteria and fungi.

REFERENCES

1. Alexander M. Biodegradation and Bioremediation. Academic Press: New York; 1999.
2. Lily Y, Young LY, Cerniglia CE. Microbial Transformation and Degradation of Toxic Organic Chemicals. Wiley-Liss Inc. New York, NY; 1995.
3. Holm-Nielsen JB, Al Seadi T, Oleskowicz-Popiel P. The future of anaerobic digestion and biogas utilization. Bioresource Technology 2009; 100: 5478–5484.
4. Steinfeld H, Gerber P, Wasenaar T, Castel V, Rosales M, de Haan C. Livestock's long shadow. Environmental Issues and Options. Food and Agriculture Organization (FAO) of United Nations; 2006.
5. Nielsen LH, Hjort-Gregersen K, Thygesen P, Christensen J. Samfundsøkonomiske analyser af biogasfllesanlg. Rapport 136; 2002. Fødevareøkonomisk Institut, København (Summary in English).
6. http://www.unesco.org/new/en/natural-sciences/ (accessed December 2012)
7. Van Agteren MH, Keuning S, Janssen D, Handbook on Biodegradation and Biological Treatment of Hazardous Organic Compounds. Kluwer, Dordrecht, The Netherlands; 1998.
8. Tokiwa Y, Calabia BP, Ugwu CU, Aiba S. Biodegradability of Plastics. International Journal of Molecular Science 2009; 10: 3722–3742.

9. Griffin GJL. Chemistry and Technology of Biodegradable Polymers, Springer, London; 1994.
10. Schmidt M., editor. Synthetic Biology: Industrial and Environmental Applications, Wiley-Blackwell; 2012.
11. NIIR Board of Consultants and Engineers (Ed.), Medical, Municipal and Plastic Waste Management Handbook. National Institute of Industrial Research, New Delhi; 2009.
12. Stuart PR, El-Halwagi MM., editor. Integrated Biorefineries: Design, Analysis and Optimization, CRC Press; 2012.
13. Perez JJ, Munoz-Dorado J, de la Rubia TJ, Martinez J. Biodegradation and biological treatments of cellulose, hemicelluloses and lignin: an overview. International Microbiology 2002; 5: 53-63.
14. Berna JL, Cassani G, Hager CD, Rehman N, Lopez I, Schowanek D, Steber J, Taeger K, Wind T. Anaerobic Biodegradation of Surfactants-Scientific Review. Tenside Surfactants Detergents 2007; 44: 312-347.
15. Fritsche W, Hofrichter M. Aerobic Degradation by Microorganisms: Principles of Bacterial Degradation. In: Rehm HJ, Reed G, Puhler A, Stadler A. (eds.) Biotechnology, environmental processes II, vol IIb. Wiley-VCH, Weinhein. p145-167.
16. Lier JB van, Tilche A, Ahring BK, Macarie H, Moletta R, Dohanyos M, Hulshoff Pol LW, Lens P, Verstraete W. New perspectives in anaerobic digestion. Water Science and Technology 2001; 43(1): 1-18.
17. Rozzi A, Remigi E. Anaerobic biodegradability: Conference Proceeding. In: 9th World Congress, Anaerobic digestion 2001, Workshop 3 Harmonisation of anaerobic activity and biodegradation assays, 9-2-2001, Belgium.
18. Dolfing J, Bloemen GBM. Activity measurement as a tool to characterize the microbial composition of methanogenic environments. Journal of Microbiological Methods 1985; 4: 1-12.
19. Soto M, Mendez R, Lema JM. Methanogenic activity tests. Theoretical basis and experimental setup. Water Research 1993; 27: 850–857.
20. Angelidaki I, Ahring BK. Thermophilic anaerobic digestion of livestock waste: the effect of ammonia. Applied Microbiology Biotechnology 1993; 38: 560–564.

21. Angelidaki I. Sanders W. Assessment of the anaerobic biodegradability of macropollutants. Reviews in Environmental Science and Biotechnology 2004; 3: 117–129.
22. Holm-Nielsen JB, Oleskowicz-Popiel P, 2007. The future of biogas in Europe: Visions and targets until 2020. In: Proceedings: European Biogas Workshop-Intelligent Energy Europe, 14–16 June 2007, Esbjerg, Denmark. Mata-Alvarez J, Macé S, Llabrés P, Anaerobic digestion of organic solid wastes. An overview of research achievements and perspectives. Bioresource Technology 2000; 74: 3-16.
23. Reusser S, Zelinka G. Laboratory-scale comparison of anaerobic-digestion alternatives. Water Environment Research 2004; 76(4): 360-379.
24. David C, Inman DC. Comparative studies of alternative anaerobic digestion technologies. M.S. (Environ. Eng.) Thesis, Virginia Polytechnic Institute and State University; 2004.
25. Riggle D. Acceptance improves for large-scale anaerobic digestion. Biocycle 1998; 39 (6): 51-55.
26. Mata-Alvarez, J., Tilche, A., Cecchi, F., editor. The treatment of grey and mixed solid waste by means of anaerobic digestion: future developments. Proceedings of the Second International Symposium on Anaerobic Digestion of Solid Wastes, Barcelona, vol. 2. Graphiques 92, 15-18 June, 1999. p302-305.
27. Christiansen N. Hendriksen HV, Jarviene KT, Ahring B. Degradation of chlorinated aromatic compounds in UASB reactors. Water Science and Technology 1999; 31: 249-259.
28. Man AWA de, Last ARM van der, Lettinga G. The use of EGSB and UASB anaerobic systems or low strength soluble and complex wastewaters at temperatures ranging from 8 to 30°C. Proceedings of the 5th International Symposium on Anaerobic Digestion. Bologna, Italy, 1988. p197-211.
29. Driessen WJBM, Habets LHA, Groeneveld N. New developments in the design of Upflow Anaerobic Sludge Bed reactors. 2nd Specialised IAWQ conference on Pretreatment of Industrial Wastewaters, October 16-18,1996.
30. Zoutberg GR, Been P de. The biobed EGSB (Expanded Granular Sludge Bed) systems covers short comings of the upflow anaerobic

sludge blanket reactors in the chemical industry. Water Science and Technology 1997; 35(10): 183-188.
31. Collins AG, Theis TL, Kilambi S, He L, Pavlostathis SG. Anaerobic treatment of low strength domestic wastewater using an anaerobic expanded bed reactor. Journal of Environmental Engineering 1998: 652-659.
32. Nagano A, Arikawa E, Kobayashi H. Treatment of liquor wastewater containing high strength suspended solids by membrane bioreactor system. Water Science and Technology 1992; 26(3-4): 887-895.
33. Garuti G, Dohanyos M, Tilche A. Anaerobic-aerobic wastewater treatment system suitable for variable population in coastal are: the ANANOX process. Water Science and Technology 1992; 25(12):185-195.
34. Oude Elferink SJWH, Visser A, Hulshoff Pol LW, Stams AJM. Sulfate reduction in methanogenic bioreactors. FEMSMicrobiology Reviews 1994; 15: 119-136.
35. Boopathy R, Kulpa CF, Manning J. Anaerobic biodegradation of explosives and related compounds by sulfate-reducing and methanogenic bacteria: a review. Bioresource Technoology 1998; 63(1): 81-89.
36. Houten RT van, Lettinga G. Biological sulphate reduction with synthesis gas: microbiology and technology. In: Wijffels RH., Buitelaar RM., Bucke C., Tramper J. (eds.) Progress in Biotechnology. Elsevier, Amsterdam, The Netherlands; 1996. pp. 793-799..
37. Jetten MSM, Strous M, Pas-Schoonen KT Van de, Schalk J, Van Dongen UGJM, Van De Graaf AA, Logemann S, Muyzer G, Van Loosdrecht MCM, Kuenen JG. The anaerobic oxidation of ammonium. FEMS Microbiology Reviews 1999; 22: 421-437.
38. Lier JB van, Lettinga, G. Appropriate technologies for effective management of industrial and domestic wastewaters: the decentralised approach. Water Science and Technology 1999; 40 (7): 171-183.
39. Abrahamsson K, Klick S. Degradation of Halogenated Phenols in Anoxic Marine Sediments. Marine Pollution Bulletin 1991; 22: 227-233.

40. Suflita JM, Horowitz A, Shelton DR, Tiedje JM. Dehalogenation: A Novel Pathway for the Anaerobic Biodegradation of Haloaromatic Compounds. Science 1982; 218: 1115-1117.
41. Annachhatre AP, Gheewala SH. Biodegradation of Chlorinated Phenolic Compounds. Biotechnology Advances 1996; 14(1): 35-56.
42. Howard PH. Handbook of Environmental Degradation Rates. Lewis Publishers: Chelsea MI; 1991.
43. Zobell CE. Action of microorganisms on hydrocarbons. Bacteriological Reviews 1946; 10(1-2): 1-49.
44. Atlas RM. Microbial degradation of petroleum hydrocarbons: an environmental perspective. Microbiological Reviews 1981; 45(1): 180–209.
45. Wilson JT, Leach LE, Henson M, Jones JN. In situ biorestoration as a ground water remediation technique. Ground Water Monitoring Review 1986; 6: 56–64.
46. Leahy JG, Colwell RR. Microbial degradation of hydrocarbons in the environment. Microbiological Reviews 1990; 54(3): 305–315.
47. Bedient PB, Rifai HS, Newell CJ. Ground Water Contamination: Transport and Remediation. PTR Prentice-Hall Inc. Englewood Cliffs, NJ; 1994.
48. EPA (Environmental Protection Agency). Monitored Natural Attenuation of Petroleum Hydrocarbons. Remedial Technology Fact Sheet. EPA/600/F-98/021. Office of Research and Development, Washington, DC; May 1999.
49. DeVaull G. Indoor vapor intrusion with oxygen-limited biodegradation for a subsurface gasoline source. Environmental Science and Technology 2007; 41(9): 3241–3248.
50. Roggemans S, Bruce CL, Johnson PC. Vadose Zone Natural Attenuation of Hydrocarbon Vapors: An Empirical Assessment of Soil Gas Vertical Profile Data. API Technical Bulletin No. 15. American Petroleum Institute, Washington, DC, 2002.
51. http://apiep.api.org/environment (accessed December 2012)
52. EPA (Environmental Protection Agency). Petroleum Hydrocarbons And Chlorinated Hydrocarbons Differ In Their Potential For Vapor Intrusion. Office of Underground Storage Tanks, Washington, D.C. 20460. 2011. p1-11.

53. www.epa.gov/oust (accessed December 2012)
54. Belay N, Daniels L. Production of Ethane, Ethylene, and Acetylene from Halogenated Hydrocarbons by Methanogenic Bacteria. Applied Environmental Microbiology 1987; 53(7): 1604-1610.
55. de Bruin WP, Kotterman MJJ, Posthumus MA, Schraa G, Zehnder AJB. Complete Biological Reductive Transformation of Tetrachloroethene to Ethane. Applied Environmental Microbiology 1992; 58(6): 1996-2000.
56. Bradley PM, Chapelle FH. Methane as a Product of Chloroethene Biodegradation under Methanogenic Conditions. Environmental Science Technology 1999; 33(4): 653-656.
57. Vogel TM, McCarty PL. Biotransformation of Tetrachloroethylene to Trichloroethylene, Dichloroethylene, Vinyl Chloride, and Carbon Dioxide under Methanogenic Conditions. Applied Environmental Microbiology 1985; 49: 1080-1083.
58. Little, CD, Palumbo AV, Herbes SE, Lidstrom ME, Tyndall RL, Gilmer PJ. Trichloroethylene Biodegradation by a Methane-Oxidizing Bacterium. Applied Environmental Microbiology 1988; 54(4): 951-956.
59. Tsien H, Brusseau GA, Hanson RS, Wackett LP. Biodegradation of Trichloroethylene by Methylosinus trichosporium OB3b. Applied Environmental Microbiology 1989; 55(12): 3155-3161.
60. Wilson JT, Wilson BH. Biotransformation of Trichloroethylene in Soil. Applied Environmental Microbiology 1985; 49(1): 242-243.
61. Pfaender FK. Biological Transformations of Volatile Organic Compounds in Groundwater. In: Ram NM, Christman RF, Cantor KP (eds.) Significance and Treatment of Volatile Organic Compounds in Water Supplies. Lewis Publishers: Chelsea, MI 1990. p205–226.
62. Bouwer EJ. Bioremediation of Organic Contaminants in the Subsurface. In: Mitchell R. (Eds.) Environmental Microbiology. John Wiley & Sons: New York 1992. p287–318.
63. Lorah MM, Olsen LD, Capone DG, Baker JE. Biodegradation of Trichloroethylene and Its Anaerobic Daughter Products in Freshwater Wetland Sediments. Bioremediation Journal 2001; 5(2): 101–118.

64. Fetzner S. Bacterial degradation of pyridine, indole, quinoline, and their derivatives under different redox conditions. Applied Environmental Microbiology 1998; 49: 237–250.
65. Shukla OP. Microbial transformation of quinoline by a Pseudomonas species. Applied Environmental Microbiology 1986; 51: 1332–1442.
66. Kaiser JP, Feng YC, Bollag JM. Microbial metabolism of pyridine, quinoline, acridine, and their derivatives under aerobic and anaerobic conditions. Microbiological Reviews 1996; 60: 483–498.
67. Carl B, Arnold A, Hauer B, Fetzner S. Sequence and transcriptional analysis of a gene cluster of Pseudomonas putida 86 involved in quinoline degradation. Gene 2004; 331: 177–188.
68. Kilbane JJ, Ranganathan R, Cleveland L, Kayser KJ, Ribiero C, Linhares MM, Selective removal of nitrogen from quinoline and petroleum by Pseudomonas ayucida IGTN9m. Applied Environmental Microbiology 2000; 66: 688–693.
69. Shukla OP. Microbiological degradation of quinoline by Pseudomonas stutzeri: the coumarin pathway of quinoline catabolism. Microbios 1989; 59: 47–63.
70. Shukla OP. Microbiological transformation of quinoline by Pseudomonas sp. Applied Environmental Microbiology 1986; 51: 1332-1442.
71. O'Loughlin EJ, Kehrmeyer SR, Sims GK, Isolation, characterization, and substrate utilization of a quinoline-degrading bacterium. International Biodeterioration and Biodegradation 1996; 38: 107–118.
72. Sugaya K, Nakayama O, Hinata N, Kamekura K, Ito A, Yamagiwa K, Ohkawa A. Biodegradation of quinoline in crude oil. Journal of Chemical Technology Biotechnology 2001; 76: 603–611.
73. Sun Q, Bai Y, Zhao C, Xiao Y, Wen D, Tang X. Aerobic biodegradation characteristics and metabolic products of quinoline by a Pseudomonas strain. Bioresource Technology 2009; 100: 5030-5036.
74. Annadurai G, Juang R, Lee DJ. Microbial degradation of phenol using mixed liquors of Pseudomonas putida and activated sludge. Waste Manage 2002; 22: 703–710.

75. Mohan D, Chander S. Single component and multi-component adsorption of phenols by activated carbons. Colloids and Surfaces A: Physicochemical &. Engineering Aspects 2001; 177: 183–196.
76. Dursun G, Cicek HC, Dursun AY. Adsorption of phenol from aqueous solution by using carbonised beet pulp. Journal of Hazardous Materials B 2005; 125: 175–182.
77. Patterson JF. Industrial Wastewater Treatment Technology, Second ed., Butterworths, London, 1985.
78. Tepe O, Dursun AY. Combined effects of external mass transfer and biodegradation rates on removal of phenol by immobilized Ralstonia eutropha in a packed bed reactor. Journal of Hazardous Materials 2008; 151: 9-16.
79. Knoll G, Winter J. Anaerobic degradation of phenol in sewage sludge: benzoate formation from phenol and carbon dioxide in the presence of hydrogen. Applied Environmental Microbiology 1987; 25(4): 384–391.
80. El-Naas MH, Al-Muhtaseb SA, Makhlouf S. Biodegradation of phenol by Pseudomonas putida immobilized in polyvinyl alcohol (PVA) gel. Journal of Hazardous Materials 2009; 164: 720–725.
81. Carrera J, Martín-Hernández M, Suárez-Ojeda ME, Pérez J. Modelling the pH dependence of the kinetics of aerobic p-nitrophenol biodegradation. Journal of Hazardous Materials 2011; 186: 1947–1953.
82. Ye J, Singh A, Ward O. Biodegradation of nitroaromatics and other nitrogen containing xenobiotics. World Journal Microbiology Biotechnology 2004; 20: 117–135.
83. Abrahamsson K, Klick S. Degradation of Halogenated Phenols in Anoxic Marine Sediments. Marine Pollution Bulletin 1991; 22: 227-233.
84. Hakulinen R, Woods S, Ferguson J, Benjamin M. The Role of Facultative Anaerobic Microorganisms in Anaerobic Biodegradation of Chlorophenols. Water Science & Technology 1985; 17: 289-301.
85. Jain V, Bhattacharya SK, Uberoi V. Degradation of 2,4-Dichlorophenol in Methanogenic Systems. Environmental Technology 1994; 15: 577-584.

86. Leuenberger C, Giger W, Coney R, Graydon JW, Molnar-Kubica E. Persistent Chemicals in Pulp Mill Effluents. Water Research 1985; 19: 885-894.
87. Sierra-Alvarez R, Field JA, Kortekaas S, Lettinga G. Overview of the Anaerobic Toxicity caused by Organic Forest Industry Wastewater Pollutants. Water Science Technology 1994; 29: 353-363.
88. Wood JM. Chlorinated Hydrocarbons: Oxidation in the Biosphere. Environmental Science & Technology 1982: 16: 291A-297A.
89. Annachhatre AP, Gheewala SH. Biodegradation of Chlorinated Phenolic Compounds. Biotechnology Advances 1996; 14 (1): 35-56.
90. Cheremisinoff NP. Biological Degradation of hazardous Waste. In: Biotechnology for Waste and Wastewater Treatment. Noyes Publications: Westwood, New Jersey, USA. 1996. p37-110.
91. Smith MJ, Lethbrideg G, Burns RG. Bioavailability and biodegradation of polycyclic aromatic hydrocarbons in soils. FEMS Microbiology Letters 1997; 152: 141–147.
92. Yuan SY, Wei SH, Chang BV. Biodegradation of polycyclic aromatic hydrocarbons by a mixed culture. Chemosphere 2002; 41: 1463–1468.
93. Ulrici W. Contaminated soil areas, different countries and contaminants, monitoring of contaminants, In: Rehm HJ., Reed G., Puhler A., Stadler P. (Eds.) Environmental Processes II Soil Decontamination Biotechnology: A Multi Volume Comprehensive Treatise, In: J. Klein (Ed.), Second Ed., vol. 11b, Wiley–VCH,Weihheim, FRG, 2000. p5-42.
94. Odokuma LO, Dickson AA, Bioremediation of a crude oil polluted tropical rain forest soil. Global Journal of Environmental Science 2003; 2: 29–40.
95. Cheung KC, Zhang JY, Deng HH, Ou YK, Leung HM, Wu SC, Wong MH. Interaction of higher plant (jute), electrofused bacteria and mycorrhiza on anthracene biodegradation. Bioresource Technology 2008; 99: 2148–2155.
96. Somtrakoon K, Suanjit S, Pokethitiyook P, Kruatrachue M, Lee H, Upatham S. Enhanced biodegradation of anthracene in acidic soil by inoculated Burkholderia sp. VUN10013. Current Microbiology 2008; 57: 102–107.

97. Li X, Lin X, Li P, Liu W, Wang L, Ma F, Chukwuka KS. Biodegradation of the low concentration of polycyclic aromatic hydrocarbons in soil by microbial consortium during incubation. Journal of Hazardous Materials 2009; 172: 601–605.
98. Santos EC, Rodrigo JS, Jacques Bento FM, Peralba MDCR, Selbach PA, Enilso LSS, Camargo FAO. Anthracene biodegradation and surface activity by an iron-stimulated Pseudomonas sp. Bioresource Technology 2008; 99: 2644–2649.
99. Zeinali M, Vossoughi M, Ardestani SK. Naphthalene metabolism in Nocardia otitidiscaviarum strain TSH1, a moderately thermophilic microorganism. Chemosphere 2008; 72: 905–909.
100. Hwang G, Park SR, Lee CH, Ahn IS, Yoon YJ, Mhin BJ. Influence of naphthalene biodegradation on the adhesion of Pseudomonas putida NCIB 9816-4 to a naphthalene-contaminated soil. Journal of Hazardous Materials 2009; 171: 491–493.
101. Gennaro PD, Rescalli E, Galli E, Sello G, Bestetti G, Characterization of Rhodococcus opacus R7, a strain able to degrade naphthalene and o-xylene isolated from a polycyclic aromatic hydrocarbon-contaminated soil. Research in Microbiology 2001; 152: 641–651.
102. Calvo C, Toledo FL, González-López J. Surfactant activity of a naphthalene degrading Bacillus pumilus strain isolated from oil sludge. Journal of Biotechnology 2004; 109: 255–262.
103. Kelley I, Freeman JP, Evans FE, Cerniglia CE. Identification of metabolites from degradation of naphthalene by a Mycobacterium sp. Biodegradation 1990; 1: 283–290.
104. Lin C, Gan L, Chen ZL. Biodegradation of naphthalene by strain Bacillus fusiformis (BFN). Journal of Hazardous Materials 2010; 182: 771–777.
105. Staples CA, Peterson DR, Parkerton TF, Adams WJ. The environmental fate of phthalate esters: a literature review. Chemosphere 1997; 35: 667–749.
106. Nalli S, Cooper DG, Nicell JA. Biodegradation of plasticizers by Rhodococcus rhodochrous. Biodegradation 2002; 13: 343–352.
107. Nalli S, Cooper DG, Nicell JA. Metabolites from the biodegradation of di-ester plasticizers by Rhodococcus rhodochrous. Science of the Total Environment Journal 2006; 366: 286–294.

108. Rahman M, Brazel CS. The plasticizer market: an assessment of traditional plasticizers and research trends to meet new challenges. Progress in Polymer Science 2004; 29: 1223–1248.
109. Tickner JA, Schettler T, Guidotti T, McCally M, Rossi M. Health risks posed by use of di-2-ethylhexyl phthalate (DEHP) in PVC medical devices: a critical review. American Journal of Industrial Medicine 2001; 39: 100–111.
110. Onorato TM, Brown PW, Morris P. Mono-(2-ethylhexyl)phthalate increase spermatocyte mitochondrial peroxiredoxin 3 and cyclooxygenase 2. Journal of Andrology 2008; 29: 293–303.
111. Horn O, Nalli S, Cooper DG, Nicell JA. Plasticizer metabolites in the environment. Water Research 2004; 38: 3693–3698.
112. Nalli SS, Horn OJ, Grochowalski AR, Cooper DG, Nicell JA. Origin of 2- ethylhexanol as a VOC. Environmental Pollution Journal 2006; 140: 181–185.
113. Barnabé S, Beauchesne I, Cooper DG, Nicell JA. Plasticizers and their degradation products in the process streams of a large urban physicochemical sewage treatment plant. Water Research 2008; 42: 153–162.
114. Beauchesne I, Barnabé S, Cooper DG, Nicell JA. Plasticizers and related toxic degradation products in wastewater sludges. Water Science & Technology 2008; 57: 367–374.
115. Deligio T. Phthalate Alternative Recognized by ECHA. 2009.
116. http://www. plasticstoday.com/articles/phthalate-alternative-recognized-echa/ (accessed December 2012)
117. Gartshore J, Cooper DG, Nicell JA. Biodegradation of plasticizers by Rhodotorula Rubra. Environmental Toxicology & Chemistry 2003; 22: 1244–1251.
118. Pour AK, Cooper DG, Mamer OA, Maric M, Nicell JA. Mechanism of biodegradation of dibenzoate plasticizers. Chemosphere 2009; 77: 258–263.
119. Firlotte N, Cooper DG, Maric M, Nicell JA. Characterization of 1,5-pentanediol dibenzoate as a potential green plasticizer for poly(vinyl chloride). Journal of Vinyl Additive Technology 2009; 15: 99–107.
120. Pour AK, Mamer OA, Cooper DG, Maric M, Nicell JA. Metabolites from the biodegradation of 1,6-hexanediol dibenzoate, a potential

green plasticizer, by Rhodococcus rhodochrous. Journal of Mass Spectrometry 2009; 44: 662–671.
121. Pour AK, Roy R, Coopera DG, Maric M, Nicell JA. Biodegradation kinetics of dibenzoate plasticizers and their metabolites. Biochemical Engineering Journal 2013; 70: 35-45.
122. http://ec.europa.eu/environment/waste/framework/index.htm (accessed December 2012)
123. Shimao M. Biodegradation of plastics. Current Opinion in Biotechnology 2001; 12: 242–247.
124. Chiellini E, Corti A, D'Antone S, Solaro R. Biodegradation of poly(vinylalcohol) based materials. Progress in Polymer Science 2003; 28: 963-1014.
125. Sakai K, Hamada N, Watanabe Y. Studies on the poly(vinyl alcohol)-degrading enzyme. Part VI. Degradation mechanism of poly(vinyl alcohol) by successive reactions of secondary alcohol oxidase and β-diketone hydrolase from Pseudomonas sp. Agricultural & Biological Chemistry 1986; 50: 989-996.
126. Suzuki T. Degradation of poly(vinyl alcohol) by microorganisms. Journal of Applied Polymer Science Applied Polymer Symposium 1979; 35: 431-437.
127. Hatanaka T, Kawahara T, Asahi N, Tsuji M. Effects of the structure of poly(vinyl alcohol) on the dehydrogenation reaction by poly(vinyl alcohol) dehydrogenase from Pseudomonas sp. 113P3. Bioscience Biotechnology Biochemistry 1995; 59: 1229-1231.
128. Bloembergen S, David J, Geyer D, Gustafson A, Snook J, Narayan R. Biodegradation and composting studies of polymeric materials. In: Doi Y, Fukuda K. (Eds.) Biodegradable plastics and polymers. Amsterdam: Elsevier; 1994. p601-609.
129. David C, De Kesel C, Lefebvre F, Weiland M. The biodegradation of polymers: recent results. Angewandte Makromolekulare Chemie 1994; 216: 21-35.
130. Chiellini E, Corti A, Sarto GD, D'Antone S. Oxo-biodegradable polymers e Effect of hydrolysis degree on biodegradation behaviour of poly(vinyl alcohol). Polymer Degradation and Stability 2006; 91: 3397-3406.
131. Khanna S, Srivastava AK, Recent Advances in microbial polyhydroxyalkanoates. Process Biochemistry 2005; 40: 607-619.

132. Ghatnekar MS, Pai JS, Ganesh M. Production and recovery of poly-3-hydroxybutyrate from Methylobacterium sp.V49. Journal of Chemical Technology and Biotechnology 2002; 77: 444-448.
133. DeMarco S. Advances in polyhydroxyalkanoate production in bacteria for biodegradable plastics. MMG 445. Basic Biotechnology eJournal 2005; 1: 1-4.
134. Mudgal S, Lyons L, Bain, J. Plastic Waste in the Environment – Final Report for European Commission DG Environment. BioIntelligence Service; 2010. http://www.ec.europa.eu/environment/ (accessed December 2012)
135. Hopewell J, Dvorak R, Kosior E. Plastics recycling: challenges and opportunities. Philosophical Transactions of the Royal Society B 2009; 364: 2115-2126.
136. Plastics Europe. An analysis of European Plastics production, demand and recovery for 2009. Plastics - the Facts 2010.
137. http://www.plasticseurope.org/ (accessed December 2012)
138. The Encyclopedia of Polymer Science and Technology, 4th Edition, John Wiley and Sons, New York; 2012.
139. Selke SE. Plastics recycling In: Harper CA. (Ed.), Handbook of plastics, elastomers and composites, 4th edition, McGraw-Hill, New York; 2002. p693–757.
140. Fact sheet, Recycling the hard stuff. U.S. Environmental Protection Agency, Solid Waste and Emergency Response, 2002 EPA 530-F-02-023 Washington, D.C. http://www.docstoc.com/docs (accessed December 2012)
141. Hendrickson CT, Matthews DH, Ashe M, Jaramillo P, McMichael FC. Reducing environmental burdens of solid-state lighting through end-of-life design. Environmental Research Letters 5. 2010. Doi: 10.1088/1748-9326/5/1/014016.
142. Cerdan C, Gazulla C, Raugei M, Martinez E, Fullana-i-Palmer P. Proposal for new quantitative eco-design indicators: a first case study. Journal of Cleaner Production 2009; 17: 1638-1643.
143. Plastic Waste: Redesign and Biodegradability. Science for Environmental Policy, Future Brief, 2001; 1: 1-8.
144. Tokiwa Y, Calabia BP, Ugwu CU, Aiba S. Biodegradability of Plastics. International Journal of Molecular Science 2009; 10: 3722–3742.

CHAPTER 7

Nilanjana Das and Preethy Chandran, "Microbial Degradation of Petroleum Hydrocarbon Contaminants: An Overview," Biotechnology Research International, vol. 2011, Article ID 941810, 13 pages, 2011. doi:10.4061/2011/941810.

CHAPTER 8

Lijun You, Kunlin Xue, Yili Kang, Yi Liao, and Lie Kong, "Pore Structure and Limit Pressure of Gas Slippage Effect in Tight Sandstone," The Scientific World Journal, vol. 2013, Article ID 572140, 7 pages, 2013, doi:10.1155/2013/572140.

CHAPTER 9

Ganapati D. Yadav and Jyoti B. Sontakke (2013). Methods for Separation, Recycling and Reuse of Biodegradation Products, Biodegradation - Engineering and Technology, Dr. Rolando Chamy (Ed.), ISBN: 978-953-51-1153-5, InTech, DOI: 10.5772/56241.

Index

A

Acrylonitrile-butadiene styrene (ABS) 198
Amorphous organic matter (AOM) 3
Argonne National Laboratory (ANL) 198

B

Biodegradation 117, 118, 125, 136, 141, 142, 144, 145
body-of-proof (BP) 49
Body-of-proof (BP) 49, 50

C

Carbon source 178
Carbon steel 42
Cellular respiration process (CSP) 177
China Automotive Technology and Research Center (CATARC) 59
Chlorinated hydrocarbon (CHC) 182
Compressed natural gas (CNG 56
Crystal nucleus growth 102, 103

D

Drainage gas recovery 23

E

Economical proces 179

Expanded granular sludge bed (EGSB) 181

G

Gaseous water 77, 78, 79, 82, 84, 85, 86, 87, 88, 93, 96
Gas slip effect 149, 150, 152
Gas-to-liquid- (GTL-) 56
Genetically engineered microorganisms (GEMs) 133
Genetically modified (GM) 135
Geochemical data 2
Good economic 24
Greenhouse gas (GHG) 55, 59

H

High-density polyethylene (HDPE) 197
High-impact polystyrene (HIPS) 198
High-performance ion-exchange chromatography (HPIC) 84
Hydrate formation 102, 103, 104, 105, 107, 108, 109, 110, 112, 113, 114
Hydroquinone (HQ) 185

I

Industrial painting 43

L

Large number 178, 182
Life-cycle analyses (LCAs) 59
Life-cycle emission model (LEM) 59
Liquefied natural gas (LNG) 56

M

Marine Corps Air Ground Combat Center (MCAGCC) 130
Methane monooxygenase (MMO) 183
Microorganism 171, 172, 173, 176, 177, 178, 183, 187, 189, 190, 195, 201, 206, 213

N

Natural gas 24, 26, 27, 102, 103, 108, 113
Nutrient 121

O

Oil density of synthetic 84
Organic fraction of municipal solid waste (OFMSW) 181
Organic matter (OM) 18

P

Pentachlorophenol (PCP) 186
Perchloroethylene (PCE) 182
Petrographic analyse 3
Petroleum-based product 116
Petroleum hydrocarbons (PHCs) 182
Physical value 35
Plasticizer 188, 212
P-Nitrophenol (PNP) 185
Polychlorinated biphenyls (PCBs) 176
Polyethylene terephthalate (PET) 194, 199
Polyhydroxyalkanoates (PHAs) 190
Polylactic acid (PLA) 195
Polyvinyl alcohol (PVA) 190
Polyvinyl chloride (PVC) 200
Precipitate lag 79

Production Index (PI) 8
Pump to wheels (PTW) 64

R

Recovery Plastics International (RPI) 197, 198

S

Software applicant 36
Solids retention time (SRT) 180, 181

T

Tetrachloroethylene (TCE) 183
Total organic carbon (TOC) 7
Total petroleum hydrocarbon (TPH) 132
Trichloroethane (TCA) 182
Trichloroethylene (TCE) 182

Tsinghua life-cycle analysis model (TLCAM) 55, 60

U

Upflow anaerobic sludge blanket (UASB) 181

V

Vinyl chloride (VC) 183
Volatile organic compounds (VOCs) 170
Volatile solids reduction (VSR) 180

W

Wastewater 179, 180, 181, 184, 192, 205, 212
Well to pump (WTP) 64
Well-to-wheels (WTW) 56, 60